Sailing Across a Wounded Sea

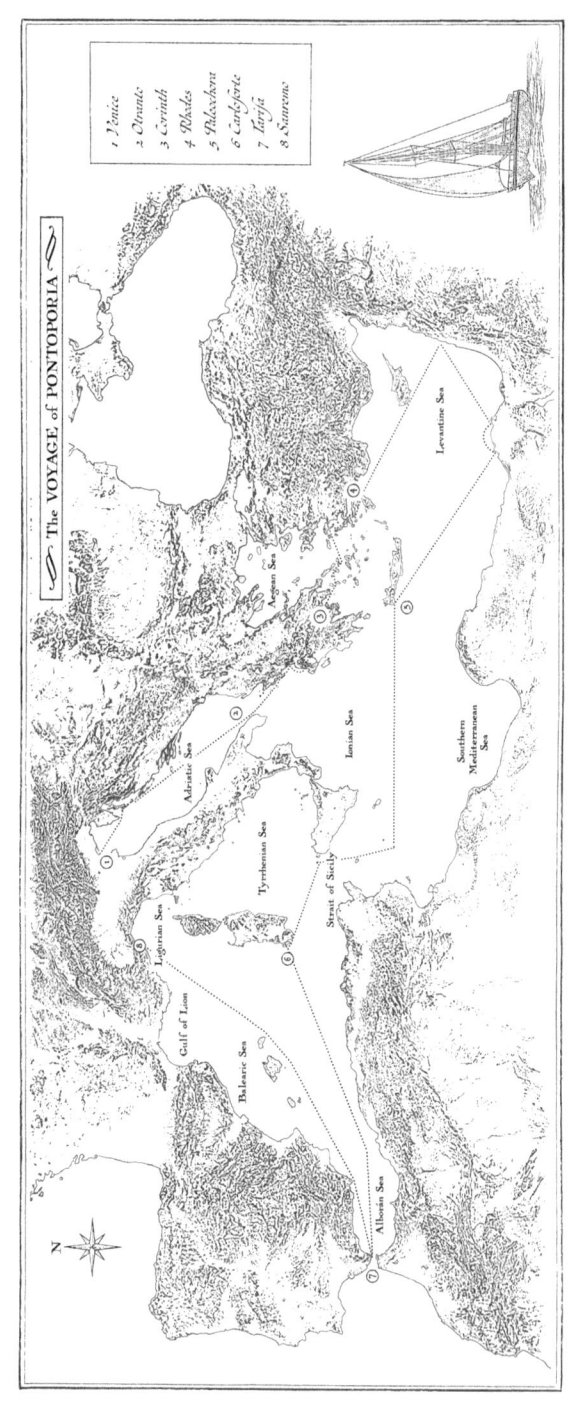

The VOYAGE of PONTOPORIA

1 *Venice*
2 *Otranto*
3 *Corinth*
4 *Rhodes*
5 *Palesthrea*
6 *Carlsforte*
7 *Tarifa*
8 *Nauvens*

Ligurian Sea

Gulf of Lion

Balearic Sea

Tyrrhenian Sea

Adriatic Sea

Aegean Sea

Strait of Sicily

Ionian Sea

Levantine Sea

Southern
Mediterranean
Sea

Alboran Sea

Illustration by Massimo Demma

Giuseppe Notarbartolo di Sciara

Sailing Across a Wounded Sea

Springer

Giuseppe Notarbartolo di Sciara ⓘ
Milano, Italy

ISBN 978-3-031-54596-2   ISBN 978-3-031-54597-9   (eBook)
https://doi.org/10.1007/978-3-031-54597-9

Cover Image: ©Simone Panigada.

This Springer imprint is published by the registered company Springer Nature Switzerland AG
The registered company address is: Gewerbestrasse 11, 6330 Cham, Switzerland

If disposing of this product, please recycle the paper.

"Drawing from decades of first-hand experiences, in *Sailing Across a Wounded Sea* Giuseppe Notarbartolo di Sciara offers a compelling narrative that interweaves the beauty of the Mediterranean Sea's non-human inhabitants with the challenges they face due to human activity. As he reflects on his journey, the author underscores the imperative for collective action and the need for heightened awareness and proactive conservation measures to protect our oceans. '*Sailing Across a Wounded Sea*' serves not only as a captivating voyage but also as a sobering call to responsibility. Join Giuseppe Notarbartolo di Sciara on this insightful expedition and discover the profound significance of preserving our marine ecosystems for future generations."

—**Peter Thomson**, *United Nations Secretary General's Special Envoy for the Ocean*

"The Mediterranean, the cradle of civilisation, is a sea steeped in history. From Apollon and Artemis to Pythagoras and Hercules, names in this book summon up images of civilisations past and lost. The greatest Greek philosopher of them all, Aristotle, is also regarded as the father of marine biodiversity, stemming from his research conducted on Lesbos Island. And yet this ancient world, familiar to so many, is under threat. In this book Notarbartolo di Sciara takes us on a modern-day odyssey through these waters, based on over 50 years of personal observations. He reminds us of the impact of the progressive footprint of human action and endeavour, which has spread out across the whole sea, changing, and depleting its beauty and splendour. His voyage paints a picture not just of what is at stake but also of the potential the future can hold if we all work together to give nature a breathing space, to allow it to restore and recover. In our modern world, where we have become more disconnected than ever from our surroundings, this book is a rallying call to remind us that all our futures are intricately interwoven with nature, and that we damage that at our peril."

—**Professor Dan Laffoley**, *Emeritus marine Vice Chair, IUCN World Commission on Protected Areas*

"Who doesn't love a travelogue? The serendipitous nature of a journey, places and characters along the way, shared insights and a joyous homecoming. A key difference with *Sailing Across a Wounded Sea* is that encounters are all with marine creatures and their places in the natural world. Set in the Mediterranean, Notarbartolo di Sciara draws upon experiences from a life-long love of the sea to take us with him: cataloguing pressures and stresses on different communities of animals he knows well. Explaining all is not as it should be. Revealing that we have been unaware of impacts, or in denial, or asleep at the wheel. Personal yet profound his consistent message is that we have taken too much and respected too little. Too much food, too much space and too many liberties. Underpinning this journey is a heartfelt call to wake up and put things right, which makes for a compelling read and a new and novel insight."

—**David Johnson**, *Honorary Professor University of Edinburgh, Coordinator Global Ocean Biodiversity Initiative, Mission Blue Hope Spot Champion*

"Everyone who cares about the Mediterranean Sea should read this impassioned and insightful book. Few people know the whales, dolphins, seals and other marine wildlife struggling to survive in the 'Cradle of Civilisation'—and what needs to be done to help secure their future—better than Giuseppe Notarbartolo di Sciara."
—**Mark Carwardine,** *Bestselling author and environmentalist*

"Jump aboard the *Pontoporia* with Giuseppe Notarbartolo di Sciara. This book is a grand tour of the Mediterranean Sea made luminous through the eyes, passions, and concerns of a dedicated scientist and conservationist who has spent his life living and working in these waters. Discover the history, culture, politics and, most of all, the diverse nature of the Mediterranean. This well-written memoir offers a wonderful ride, though at times bumpy, as a scientist who has done so much to help the Mediterranean come to terms with the future of this ancient, storied sea."
—**Erich Hoyt, Author, Planktonia,** *Creatures of the Deep, Marine Protected Areas for Whales, Dolphins and Porpoises, and other books*

"Warm and inviting as a Mediterranean breeze, renowned marine biologist Giuseppe Notarbartolo di Sciara's beautiful book takes us on a modern-day Odyssey around the sea of his birth. It is a journey of entrancing encounters with the natural world, tempered by a sobering warning – that the 'Cradle of Civilisation' cannot call itself civilised if it continues to destroy the sea that gave it life."
—**Isabella Tree, Author of** *Wilding*

Prologue

I have been waiting for you. All my life I have been waiting for you, every time I found myself encroaching upon your world. Scanning the sea surface for hours, day after day, looking for tell-tale signs of your presence: a ripple in the water or the flash of a pair of flippers breaking the surface, caught with the corner of the eye. I knew you were there, somewhere. But you persisted unseen.

With the passing of years, I have learned to take solace in my search by simply knowing that you existed, by valuing the notion of your existence as a surrogate of a missing direct experience of you. Enjoying the beauty of the world you evolved to live in, so pleasant to body and mind as if my habitat were to coincide with yours. Because it does, to some extent: waters transparent like the purest of crystals; the fragrance of sun-baked rockrose, lentisk, thyme, and wild sage wafting down to the water from the scarp cliffs; the antics of screeching Eleonora's falcons soaring overhead. Basking in the blissfulness of your home had become a proxy for the excitement of seeing you in flesh and bone.

You had good reasons to be shy, for you were, and still are, the rarest of the world's seals. They called you Mediterranean monk seal, with nobody to this date able to convincingly explain the origin of that puzzling monastic attribute. Your species was not meant to be a rare species. There is nothing wrong with you and the ability of your tribe to make a decent living along the Mediterranean shores. And in fact, in past times, you populated the beaches throughout the Great Sea, from Gibraltar to Palestine and from Venice to Gabès. Records of your abundance abound, starting from Homer, that very first chronicler of the tribulations of Mediterranean men and women. In the Odyssey, the poet narrates about flocks of seals being herded by Proteus, son of Poseidon, around the island of Pharos, just off what later became the harbour of Alexandria.

Many accounts after that, across the millennia, provide unambiguous evidence of your former prosperity. Until recently, to be precise. It was mainly during the past century that the relentless persecution you received from humans has managed to push you to the brink of extinction, reducing your numbers to just a few hundred individuals, thinly spread over remaining strongholds in the Eastern Mediterranean and on a few remote Atlantic islands and shores.

The reason for so much animosity was, and still is, quite unfair. True, you do damage nets from time to time, but this is only because humans have left too few fishes in the sea for you to subsist on. Theirs is entirely the responsibility for destroying the natural balance in the marine environment, and you are a most convenient scapegoat.

So few of you exist today that the man in the street, having lost the experience and the knowledge of your kind, is mind-boggled when told of the existence of seals in this part of the world. "Seals in the Mediterranean?"—is the typical reaction—"Come on, I cannot believe it. Seals only live in cold waters". So it is that, on top of everything, you are even being deprived of the dubious benefit of being mourned for your impending demise.

You might wonder why I have been so keen on seeing you, why I developed this special relationship with an almost obsessive idea of you in my mind. It was the abstract idea of the monk seal that I was seeking. Mediterranean monk seals have become a symbol of the wilderness that my sea is losing because of the senseless destructiveness of my species. Touching your continued existence with hand would have worked as a reassuring charm, rescuing me from a sentiment of distress.

And then, on that mid-summer afternoon, you appeared to me when I least expected to see you. I had not given up on finding you, of course; I was only momentarily distracted. I was sitting on my caique reading a book, lulled by a gentle swell, anchored in the shade of a cliff on the island of Patmos. You surfaced in front of me, your large round head silently and slowly emerging from the water like a magic apparition, with your long, dripping whiskers and a small, remarkable white spot between your eyes. What I remember most distinctly was your gaze: your big, dark eyes staring at me, straight into mine, piercingly. Eyes that were not smiling (Fig. 1).

I was spellbound. After my long search, there you were, but only when you had decided to reveal yourself—how very fittingly, right in the shade of the island of the Apocalypse, or Revelation.

What I had in front of me was not the abstract idea of a monk seal but the real thing: breathing, staring at me in direct, person-to-person contact. It was clear that it was you, not me, in the full faculties of your phocine personhood, who called the shots.

Fig. 1 "You surfaced in front of me, your large round head silently and slowly emerging from the water like a magic apparition, with your long dripping whiskers and a small, remarkable white spot between your eyes" (Illustration by Massimo Demma)

I was struck by the painful asymmetry of the situation. You were not reciprocating my affectionate emotion. You likely couldn't care less about me. After staring at me with that stern gaze that seemed to last forever, you gave a curt snort and disappeared from sight below the surface. The spell was over as you resumed doing the things that seals normally do. Minutes later, I did see your back briefly surfacing again at a distance, and then again even farther on before you were no longer visible as you patrolled the reef in search of some meagre prey, leftovers from your greedy human competitors.

There were so many things I wished I could ask you, but mostly, I wanted to tell you how appalled I was by the terrible wrong humans have been inflicting on you. At that moment, I remember a sense of outrage against my kind simmering within, and the formulation in my mind of the resolve that I would try my best to right that wrong. To do something, anything, that could contribute to improving your condition.

Patmos, Greece
1 August 2002

Acknowledgements

On 15 March 2018, as I was in Kota Kinabalu, on the island of Borneo, for a workshop to identify Important Marine Mammal Areas (IMMAs) in the North East Indian Ocean and South East Asian Seas region, I was having dinner with Erich Hoyt, long-time friend and *partner in crime* in the IMMA adventure. At one point during the dinner, Erich looked at me and said: "You know, you should write a book to tell all your Mediterranean stories". It was there and then that the spark for writing this book was ignited. Perhaps out of guilt for having sent me down a rabbit hole that kept me toiling during my spare time for the ensuing 6 years, Erich was constantly on call with all his expertise and an extra dose of patience to lend his unfortunate colleague all the advice and support he desperately needed during this writing and publishing adventure.

I am immensely grateful to Editor Eva Loerinczi, who encouraged me to submit the manuscript to Springer Nature and made its publication possible. Thanks to the competent collaboration of Bibhuti Bhusan Sharma and Manigandan Jayabalan, the book production process ended up being quick and smooth. I wish to thank Elizabeth V. Hillyer for ensuring that my writing would not turn out to be too dissonant from Shakespeare's idiom, and Massimo Demma for his artistic skill and patience in preparing the maps of *Pontoporia's* voyage and the beautiful portraits of the main marine characters that this book is about. My son Marco, a diplomat at heart and much wiser than his father, strived to redress elements of political incorrectness interspersed in my prose, occasionally irate towards the most egregious agents of environmental damage; elements that survived his revision are my responsibility alone.

My friend and colleague Sabina Airoldi gave me invaluable support by providing access to many photographs of the highest quality that inspired Massimo Demma to create this book's illustrations. These include a picture of a striped dolphin and one of a pilot whale taken by N. Pierantonio, one of a fin whale and one of a spinetail devil ray taken by L. Lodigiani, one of Risso's dolphins taken by V. Fadda, one of a sperm whale taken by G. Passoni, and one of a Cuvier's beaked whale taken by herself. Joan Gonzalvo provided, for the same purpose, pictures that he had taken of a bottlenose dolphin, a common dolphin, and a Mediterranean monk seal. The image of rough-toothed dolphins was based on underwater video footage taken by Silvia Frey in 2016, and that of orcas was inspired by photographs courtesy of Alfredo Rodrigues, "Ocean Vibes Algarve".

Simone Panigada is the author of the stunning cover photo. We were together on a dinghy, floating in the calm waters of the Pelagos Sanctuary, when a fin whale surfaced between us and our vessel. I was filming the scene, and the photo is his.

This book contains interviews of colleagues Draško Holcer, Joan Gonzalvo, Claudio Cuoghi and Spyros Kotomatas. I thank them all for revising the text portions that concern them, agreeing to their publication, and ensuring that they correctly reflected their thoughts.

My highest appreciation goes to Mark Carwardine, Hal Whitehead and Carl Safina for advising and helping me in my search for a publisher before I submitted my book proposal to Springer Nature. Thanks also to Simone Repetto for providing information on the tonnara on the Island of San Pietro in Sardinia, and to my sister-in-law Donata Pizzi for her learned advice on Dodecanese architecture.

I owe much of the way I think about nature and science to the cultural medium in which I had the extraordinary luck of growing up, and to the single persons who had made such medium substantive, rich, and meaningful. All the ideas, decisions, and initiatives I took during my lifetime are connected, one way or the other, to very special minds and characters that have inspired and supported me in many ways. First and foremost I recall my beloved mentor, the oceanographer Walter Munk, who, with his wife Judy, hosted me at Seiche, their extraordinary La Jolla residence, for almost a decade. Like a father when I was taking my first steps as a marine biologist, Walter taught me about the challenge of dealing with uncertainty in the effort to understand natural phenomena, and the art of turning setbacks into successes. Walter's second wife, Mary, supported the continuation of his moral legacy after his death. Luigi Cagnolaro, the late director of the Natural History Museum of Milano, was an important role model in his unique combination of scientific rigour, zoological passion, human kindness and hilarious humour. Tundi Agardy, that splendid, indomitable marine scientist/warrior, taught me everything there is to know about protecting ocean spaces, right in that mid-life phase in which my professional inclination was morphing from scientific curiosity to environmental commitment.

The Tethys Research Institute—funded in 1986 on the initiative of the late Egidio Gavazzi, and named after the mythological demi-goddess, wife of titan Oceanus—was the cultural incubator where all the colleagues were hatched from, who have walked with me for the good part of half a century along the most exciting pathways of marine discovery: Sabina Airoldi, Arianna Azzellino, Giovanni Bearzi, Silvia Bonizzoni, J. Fabrizio Borsani, Amina Cesario, Marina Costa, Maddalena Fumagalli, Adriana Geraci, Joan Gonzalvo, Maddalena Jahoda, Simone Panigada, Elena Politi, Margherita Zanardelli, to mention only the ones whom I worked closer with over extended periods on a day-to-day basis. Later came the exciting new (and ongoing) adventure of identifying Important Marine Mammal Areas across the world's oceans, in the intrepid company of Erich Hoyt, Gill Braulik, Gianna Minton, Randy Reeves, Vienna Eleuteri, David Johnson, Dan Laffoley, Mike Tetley—accompanied by several of the above-mentioned Tethys' colleagues—and many, many others. Throughout my lifetime, with the only notable exception of when I was entangled in the drab atmosphere of government, I enjoyed the priceless privilege of working in the company of friends.

The family ecosystem I have been lucky enough to thrive in has been fundamental to my sanity. It has all been pivoting around Flavia, my wife, our children Marco and Bianca and grandchildren, and my many sisters with their families. I could not be grateful enough to all of them for having created around me the solidest life-support system.

Last of all, and most of all, I wish to thank the individual whales, dolphins, seals, sharks, rays, turtles, fishes, and birds that I have encountered in the Mediterranean during a life spent at sea looking for them—for all the inspiration and the wild joy they unwittingly gave me every time I could make contact with them, and for having made me feel so alive in their company and welcome in their world as if I had been one of them. Because that is, in reality, what I was.

About the Book

The capacity of humans to destroy their environment is playing out like a Greek tragedy in the Mediterranean Sea. After having coexisted with a diversity of marine animals throughout their history, humans have broken the balance in recent decades, and the survival of countless marine creatures is now increasingly uncertain. However, unlike in classical tragedies, real-life entities are not necessarily doomed by their fate, and there must be hope to turn the tide in nature's favour. Lack of concrete conservation action might be simply due to a lack of awareness: how can we feel concerned about a loss if we don't know what we are losing? *Sailing Across a Wounded Sea* is the story of an ideal journey around the Mediterranean to meet its non-human inhabitants: a reconstructed collage of really happened episodes collected over half a century as the author observed real animals, exchanged views with people, and argued for such views in the policy arena.

Giuseppe Notarbartolo di Sciara has been involved for a lifetime in protecting Mediterranean marine biodiversity in various capacities—as a scientist, civil servant, advocate, and sailor. Having studied in California and worked with whales, dolphins and sharks worldwide, he returned to the Mediterranean in 1985, keen on using his acquired tools to discover more about the ancient sea's natural history. Here, he described small but vibrant populations of fin and sperm whales, along with various species of dolphins, devil rays and the monk seal.

At the same time, seeing the sea's progressive degradation at the hands of humans, he feels a surge of rebellion against this squandering of natural values and hopes that encountering whales, dolphins, seals, and rays in their habitat and on their terms will contribute to building up in readers a collective commitment to secure a future for these species. A future in which they are allowed to flourish as they were meant to, had humans never trod so heavily on the sea's delicate ecological balance and the interwoven natural processes.

Contents

About the Author

Giuseppe Notarbartolo di Sciara is a marine ecologist who spent his life trying to shoehorn scientific evidence into policy and politics. So far, he must admit that he has largely failed, despite episodical successes here and there. He is fascinated by devil rays and horrified by seafood after a childhood encounter with a trout gone bad. When he is not on the Greek island of Patmos he spends his time in Milan, thinking of ways to get back to Patmos.

List of Figures

1

Introduction: Mediterranean Stories Worth Telling

In the distant past, our planet did not look like it does today because continents were arranged differently on the world map. About 250 million years ago, they were all jumbled into a single landmass we called Pangea, surrounded by a single ocean called Tethys. But Pangea, under the impulse of continental drift, was not meant to stay as it was. The landmass started to break apart, and the pieces—the continents—moved around, progressively changing their relative positions until they reached the familiar configuration. As continents moved about, so did Tethys, filling with portions of the ocean the interstices between them, as these formed under the push of continental drift. At one point, the continents known today as Africa and Eurasia (Europe plus Asia), in their movements, pinched between them a small remnant of Tethys that became the Mediterranean Sea.

Eventually, about 6 million years ago, Africa and Eurasia moved towards each other a bit too much, cutting off the Mediterranean from the rest of the ocean. Six million years ago is a relatively recent time as geological events go, considering that Earth's age is estimated at 4.5 billion years; scaling down the lifetime of our planet to a span of 24 hours, the event in which Mediterranean became isolated would have happened no earlier than 2 min ago. In that episode, known to geologists as the Messinian Salinity Crisis, water subtracted into the atmosphere by evaporation from the Mediterranean could no longer be replaced through a connection with the rest of the ocean. This situation caused the basin to dry up within the time span of a few hundred thousand years, with only puddles of hyper-concentrated brine remaining at its bottom. Along with its waters, almost all the original Tethyan marine organisms living in the Mediterranean were wiped out. Only a few species survived, mainly from the crustacean and mollusc tribes, having been able to withstand the extreme levels of salinity and temperature resulting from the upheaval. Then, 670,000 years later, the African and Eurasian plates moved ever so slightly once more, and a connection with the world's ocean opened up again where Gibraltar is today. Seawater started rushing back into the basin with titanic impetus, the flow rate reconstructed to be a thousand times that of the Amazon River. Impetus notwithstanding, the complete refilling of the Mediterranean might have taken no less than an entire decade.

At the beginning of the Mediterranean's new lease on life, modern humans did not exist yet. It was the time in which our Hominine ancestors, all concentrated in the centre of Africa, split from Chimpanzees, and a good 2 million years before their various Australopithecine descendants started experimenting with stone tools. The world had to wait for 3 million more years for *Homo sapiens* to make its fateful appearance. Before him, thousands of animal forms of life - from bacteria to whales—had come in from the Atlantic Ocean across the Gibraltar breach to call the Mediterranean home, flourishing in its hospitable waters and balmy climate, evolving into increasingly well-adapted forms, contributing with their existence to the character and the health of their marine and coastal environment, and ultimately accruing to an unrepeatable, exquisitely balanced combination of biotic and abiotic ingredients.

When humans discovered the Mediterranean coasts during their out-of-Africa migrations, some were quick to notice the favourable environmental conditions offered by this sort of promised land, and settled in to stay. These conditions proved to be essential for their prosperity. It was not by chance that the Mediterranean is known as the cradle of Western civilisation: the region's environment was the backdrop against which human societies developed. It was a powerful ingredient in the recipe that shaped the Western world's citizens and their thinking and achievements across the millennia.

Mediterranean marine and coastal organisms were essential to the development of the region's human cultures, in part as exploitable sustenance resources and in part integrated into their myths, enshrined in local lore and protagonists of the most diverse traditions and stories, from the Bible to Greek mythology and from the Odyssey to Pinocchio. However, as soon as the newcomers started handing down stories about their adventures, the narrative became more and more absorbed in themselves, with the original Mediterranean landlords relegated to an increasingly rarefied cloud of irrelevance. None of the representatives of Mediterranean charismatic megafauna come even near to retaining for our peoples the cultural value, for instance, that dugongs have for the indigenous Australians, the whales for the Māori, the orcas for the First Nations of the Canadian Pacific, and the sharks for many Polynesians.

This drift probably occurred as the Mediterranean peoples' cultural framework turned sharply away from nature. There was a stark divergence here from philosophies from other parts of the world, such as Buddhism, Taoism, and Confucianism in Asia that recognised a universe in which all parts—humans included—are interdependent, embedded within nature, with us humans expected to be in awe, respectful and even guardians of its indivisible harmony.[1] In the Mediterranean, arising monotheist doctrines adopted the opposite approach by peremptorily placing humans on a pedestal of purported superiority over all other beings. Once appointed (self-appointed, in fact) as masters of the world, they were given a privilege from above to conquer, dominate, and exploit without any prescription for care. In their delusion of grandeur, Mediterranean humans saw little use in providing the natural world and its non-human inhabitants the attention and respect they deserved. This neglect has become so widespread across the millennia and so typical of Western thinking that it persists today as the predominant way of conceiving humans' relationship with nature.

This state of things is well exemplified in history books describing the human trajectory along the Mediterranean shores and across its waters from the beginning to the present day. Of course, when mentioning history, what is typically meant is human history, and it would be naïve to expect otherwise. And it cannot be denied that historical accounts of the human conquest of the region—e.g. by Fernand Braudel,[2] David Abulafia,[3] and Predrag Matvejevic,[4] to name three of my favourites—make for inspirational reading.

And yet, as engaging as such stories are, whilst wading through these works, I keep sensing that an essential element is missing: what about all the other Mediterranean peoples, the non-human ones that had been here long before us? Of course, there are fewer details we can delve into about the whales, the dolphins, the turtles, and the sharks than we can about humans because we know ourselves much better than we know them. Nevertheless, the little we know about them is no less interesting nor deserves less attention, if only because of the interconnections with our own story. This is why I find myself so often craving for stories that are not so monothematically human-centred, produced by an increasingly introspective and self-referenced culture that has

---

[1] Lent J. 2021. The web of meaning. Profile Books Ltd., London.
[2] Braudel F. 2009. La Méditerranée: l'espace et l'histoire. Flammarion.
[3] Abulafia D. 2014. The Great Sea: a human history of the Mediterranean. Penguin.
[4] Matvejević P. 2020. Breviario mediterraneo. Garzanti.

successfully managed to segregate the scholars of non-humanistic disciplines within the walls of dusty natural history museums or anodyne laboratories, and condemned them to mainly talk to themselves.

This gaping rift that has developed in the most learned Mediterranean public between a majority who have regard chiefly for humanities and the others who are interested in science and nature,[5] is not only unnecessary and anachronistic, for there is never a good reason for creating barriers within the realm of knowledge; it is also harmful. The evident lack of interest in the natural world affecting such a large portion of Mediterranean societies, which includes influential categories such as lawmakers, teachers, judges, writers, and the media, is at the root of our inability to grasp the importance of timely addressing environmental crises caused by humanity, such as the disruption of the planet's climate, the acidification of the oceans' waters, the depletion of biodiversity, the extinction of species, the dissemination of contaminants and plastics and the spread of pandemics, with all the tightly interrelated effects that these novel Horsemen of the Apocalypse inflict on the wellbeing of Earth's inhabitants, humans included.

Mine here is an attempt to help kindle the interest of readers in the wondrous natural aspects of the Mediterranean region by telling stories about the picturesque marine characters in the play, such as the whales, the dolphins, the seals, the turtles, the sharks, and the rays—in short, the so-called "charismatic megafauna"—which I have been so fortunate to encounter during half a century of fieldwork as a marine conservation ecologist. I know that it will come as a surprise for many to discover that the Mediterranean Sea, commonly considered by the layperson a region poorly endowed with interesting natural features, is not quite what most people think, and that the sea is populated by fascinating beings, hiding in plain sight as they struggle to survive in an environment that humans are rendering increasingly inhospitable.

Kindling interest in Mediterranean marine fauna comes none too soon. So many Mediterranean species need all the help they can get, and fast, because their survival is increasingly uncertain. And yet, surprisingly, none of the famed historical accounts mentioned above that so masterly tell the exploits of humanity in the Mediterranean across the millennia care to adequately capture the fact that this same humanity, with all its wondrous achievements in terms of philosophy, arts, music, poetry, gastronomy, farming, technology, searches for beauty and creativity, has morphed today into a monster unable

---

[5] This rift was famously lamented by British novelist and chemist Charles P. Snow, who noted in 1959 the splitting of the intellectual life of the whole of Western society into two cultures, "both sides developing their own language and ending up creating an unsurpassable cultural divide preventing each side from understanding and appreciating the other".

to avoid soiling the very cradle of the civilisation it created, and on a fast track to destroying the region's inhabitants, itself included.

The Mediterranean marine environment is known to be amongst the world's most imperilled, and this is only because of the intensity of human activities. Ninety-six percent of Mediterranean fish populations (fishery scientists call them "stocks" probably because they see them neatly arranged along the shelves of an imaginary warehouse, ready for marketing and human consumption) are being captured unsustainably, with an average ratio of current fishing three times higher than fishing at a sustainable level.[6] Plastic pollution in the Mediterranean is one of the planet's densest. The Mediterranean is also one the world's busiest waterways, sustaining one-fifth of the global maritime trade over a surface of just 1% of the world's oceans: a feat not without consequences, as marine traffic is a significant source of noise, disturbance, and pollution. Not to be outdone, the world's navies insonify Mediterranean waters with their powerful submarine warfare sonars, whilst industrial fleets instrumented with deafening contraptions complete the job in search for oil and gas hidden within the sea bottom. All this noise is having a dire effect on the acoustically hypersensitive whales and dolphins. But the straw that breaks the camel's back is the warming of the sea due to the human-caused disruption of climate, which leaves very little escape options for marine animals wishing to shift their range northward to compensate for temperature increase—an option precluded to them by the geography of Mediterranean coastlines.

And yet, the health of the Mediterranean Sea is essential not only for the non-human beings that live in it; it is substantial for us humans as well. Benefits to humans from the Mediterranean natural environment are of many types, material and immaterial—such as beauty and inspiration—but let me simply note here the economic values, a currency everybody understands well. Although, as we have seen, the Mediterranean accounts for only 1% of the world's ocean, it generates about 20% of the global revenues related to maritime activities. Today's resources produce benefits worth US$450 billion annually, supporting jobs and wellbeing for almost 500 million people in the region. This value depends to a large extent on the marine ecosystems' health and biodiversity, which is currently in dire danger.[7]

No wonder so many marine inhabitants in this region are threatened with extinction: a sad irony considering that the Mediterranean is one of the world's

---

[6] Mediterranean fish stocks on the brink: https://tinyurl.com/mrmdj9ne.

[7] Piante C., Ody D. 2015. Blue growth in the Mediterranean Sea: the challenge of good environmental status. MedTrends Project. WWF-France. 192 p.

top biodiversity hotspots.[8] In its cramped space—one-hundredth of the global ocean surface—the Mediterranean Sea hosts more than 6% of the known marine species of animals and plants, many of them endemic. And yet, this natural wealth is at risk of disappearing. One-third of the seagrass beds have gone. Most marine mammal species and more than half of the shark and ray species regularly occurring in the Mediterranean are in a threatened category in the Red List of Threatened Species compiled by the International Union for the Conservation of Nature.[9] It is a miracle if megafauna still occurs there. Its terrestrial counterparts—all those lions, leopards, hippos, and even elephants trotting along the Mediterranean shores until historical times—have gone extinct because human-induced extinction progresses faster on land than in the sea. Bears and lynxes hang on by a thread; wolves are regaining terrain because they have been strictly protected recently and because of the species' inherent resilience. Marine animals, by contrast, are more difficult to dispose of, as it takes longer to extirpate species from their watery habitats. So, they are still with us—but many of them just barely.

Lawmakers from the Mediterranean coastal nations recognised time ago that their sea required urgent political attention and resolved to express their concern in 1976, during a meeting in Barcelona, by agreeing on a treaty called the "Convention for the Protection of the Marine Environment and the Coastal Region of the Mediterranean." Beautifully crafted and assisted by competent, dedicated professionals, the Barcelona Convention remains, unfortunately, not much more than a "gentlemen's agreement"—because the pussyfooting signatories decided that respect for their collective decisions had to be based on the goodwill of the concerned instead of on clear rules. For example, endangered animal and plant species—from whales to seagrasses—are listed as protected in a special protocol of the Convention. Still, the mechanisms for ensuring the compliance of such protection are lame. Unfortunately, gentlemen's agreements require gentlemen to function, too rare a breed when managing humans' effects on the environment is concerned.

Meanwhile, the men and women in the streets in Rome, Barcelona, Athens, or Tunis today are largely oblivious to what lives, swims, and dies in the Mediterranean Sea, and ignore that marine biodiversity is slipping away from

---

[8] Coll M., Piroddi C., Steenbeek J., Kaschner K., Ben Rais Lasram F., Aguzzi J., Ballesteros E., Bianchi C.N., Corbera J., Dailianis T., Danovaro R., Estrada M., Froglia C., Galil B.S., Gasol J.M., Gertwagen R., Gil J., Guilhaumon F., Kesner-Reyes K., Kitsos M.-S., Koukouras A., Lampadariou N., Laxamana E., López-Fé de la Cuadra C.M., Lotze H.K., Martin D., Mouillot D., Oro D., Raicevich S., Rius-Barile J., Saiz-Salinas I., San Vicente C., Somot S., Templado J., Turon X., Vafidis D., Villanueva R., Voultsiadou E. 2010. The biodiversity of the Mediterranean Sea: estimates, patterns, and threats. PloS ONE 5(8):e11842. doi:10.1371/journal.pone.0011842.

[9] IUCN Red List: https://tinyurl.com/56djmsuf.

our lives really fast. An increasingly urbanised population is affected not only by the extinction of nature but also by the extinction of its own "natural experience"; and so it is that ignorance, in turn, spawns indifference. How can we care for something the existence of which we are unaware? As Senegalese conservationist Baba Dioum once said, "In the end, we will conserve only what we love, we will love only what we understand, and we will understand only what we know".

The whales, the dolphins, the seals, and the sharks encountered in these pages first occupied my scientific mind, only to move to my heart as I started to see them not so much as study subjects but as fellow passengers in the interstellar trajectory aboard spaceship Earth. Despite being all fundamental players in the balance of the marine ecosystems that are everyone's lifeline, these species persist unnoticed by human landlubbers who have lost the ability to see much of what is non-human around them. I wish that encountering these animals, albeit only through ink on paper, and pondering on their struggle, might provide the stimulus for considerations that go well beyond the mere geography of my story.

The time has come to take stock of mistakes and redress human behaviour. I do not think that humans misbehave because they are inherently evil. There is no element of misanthropy in my effort. Quite the opposite, humanity is, for me, a constant source of wonder for all its extraordinary accomplishments. However, precisely because of these accomplishments, humanity is caught in a trap, unable to shake itself from the intoxication of the power afforded by technology (with the added recent injection of support from artificial intelligence); and incapable of reining in such inebriation by imposing on itself the wisdom of restraint. If we manage to free ourselves from the trap, nature will again lend her hand as we mend a rediscovered alliance.

The elusive Mediterranean monk seals, sticking their round moustached heads out of the surface to stare at us, can be a powerful symbol of this desirable new deal, for if monk seals will be able to make a comeback from oblivion—as they seem to be doing—anything else can follow. The monk seal "revelation" I had on the island of Patmos decades ago, told in the Prologue, embodies the metaphor that illustrates the main messages I wish to convey: one of regret for what humans did to our common house and our co-tenants, of resolve to help to right the wrongs, and of respect not just for the marine species intended as Aristotelian categories—"the" monk seal, "the" fin whale, "the" loggerhead turtle—but for all the individual Mediterranean inhabitants that go under such categories: these cognitively-, emotionally- and socially-capable non-human persons who deserve our awareness, sense of justice, respect, empathy, and ultimately, love.

Often, interesting ideas come during the night, and one night, as I was tossing myself in bed between one sleep cycle and the other, I had this idea. During my lifetime, I accumulated a good deal of first-hand stories about the Mediterranean non-human inhabitants. What if I imagined telling all these stories by embarking on a 5000-mile,[10] 2-month-long journey across the Mediterranean Sea? This imaginary journey would begin in early summer 2021 from Venice, my place of origin, and end in Sanremo in the Ligurian Sea, the base my colleagues and I have been using for decades for our scientific explorations—after having sailed clockwise across all the main seas composing the Mediterranean. It would be a long voyage, "full of adventure, full of knowledge", as prescribed by Constantine Cavafy in his timeless poem "Ithaka"[11]: a journey where what matters is not the point of arrival but the journey itself, with all the wisdom derived from the experience and knowledge received along the way.

Accordingly, the story unfolding in these pages is a collage of real episodes that happened to me over a span of half a century in which I had the opportunity to sail across real waves, observe and discover real animals, exchange my views with real people, and argued for such views in the policy arena. All the episodes, deconstructed and reassembled here for a narrative purpose to create a single journey, are experiential tesserae from different places visited at different times, rearranged to create the mosaic of a single ideal journey. As a consequence, the fictional element of the story is its temporal dimension. I hope that the resulting narrative will convey to my readers the excitement of meeting the wondrous inhabitants of the Mediterranean Sea and help to build a collective wish of securing a future in which they will be left to flourish the way they were meant to, had humans not treaded so heavily on such a delicate balance of lives and natural processes.

Pontoporia,[12] the sailing boat aboard which I am making the journey, is also an imaginary vessel impersonating and reassembling in herself the essence of the many real vessels—Santal, Gemini Lab, Fling, Pelagos, Sirius, Saen, Chance—that have hosted and protected me against the sea's vagaries during my voyages, and innumerable other boats that I have ferried across from one

---

[10] Distances at sea are customarily measured in nautical miles, so they are here wherever they pertain to nautical narration. When miles are mentioned in these pages, they are always meant to be nautical miles—one nautical mile being 1,852 metres long and equivalent to one prime of Latitude. Furthermore, speed at sea is measured in "knots", i.e. the number of nautical miles covered in one hour.

[11] Constantine P. Cavafy. 1911. Ithaka. https://tinyurl.com/ykczk6st.

[12] *Pontoporia blainivillei* was also the name used by zoologists Paul Gervais and Alcide d'Orbigny in 1844 to describe the La Plata dolphin, also known as "franciscana", a dolphin found in the coastal and estuarine waters of southeastern South America (and therefore not in the Mediterranean)—the species so named to celebrate the nineteenth century's French zoologist Henri de Blainville.

port to another or chartered for various purposes during my lifetime, whose names I have long forgotten. I received the inspiration for my boat's name from Hesiod, who wrote that in Greek mythology "Pontoporia" or "Pontoporeia" ("the seafarer") was the Nereid of sea crossings, one of the 50 marine-nymph daughters of the "Old Man of the Sea" Nereus and the Oceanid Doris. Being a fictitious boat, I have endowed *Pontoporia* with all the latest technologies to help me in my solo navigation, sparing no (imaginary) expense: radar to avoid running into other ships at night or in the fog; global positioning system (GPS) to know my exact position at any time; depth sounder to avoid running onto some shoal; autopilot to relieve me from constantly staying at the wheel; bow-thruster to assist manoeuvrings in harbours; automatic identification system (AIS) to be aware of surrounding traffic; water desalinator to enjoy a constant supply of fresh water aboard; and a hydrophone (i.e. underwater microphone) installed in the hull to be able to hear the voices of the marine animals encountered. All these technological apparatuses, commonly available in my travel year, allowed me to progress comfortably and safely at sea in ways that no sailor could have imagined even a few decades ago. Finally, I hope my readers will forgive me for having told of my repeated trouble-free boundary crossings from harbour to harbour and from country to country, without wasting hours in the various customs offices. We all know that this is nothing but a fictional subterfuge, but I hope in a general agreement that adding an over-realistic picture of the actual hardship of crossing maritime borders would add nothing of interest to my narrative.

2

Venice

The window in the classroom was open, and warm springtime air was wafting in. We had been watching for weeks the old plane tree in the middle of the yard of the elementary school "Armando Diaz" becoming covered with new green leaves as the season progressed. The air was full of the cries of cavorting swifts, freshly arrived from the south of Africa, announcing that summer vacations were around the corner. Mrs. Angela, our teacher, had instructed us to use the last hour of class to make a drawing of anything we fancied. We kids loved the idea because it had been a long day, and we all welcomed an earlier break from the Italian, arithmetic, and history lessons that had kept us busy during the year.

I felt inspired. In minutes, I sketched what I had in mind and turned in my artwork to Signora Maestra. The other kids were still intent on their drawings, so there was time for her to look at my production and comment (Fig. 2.1).

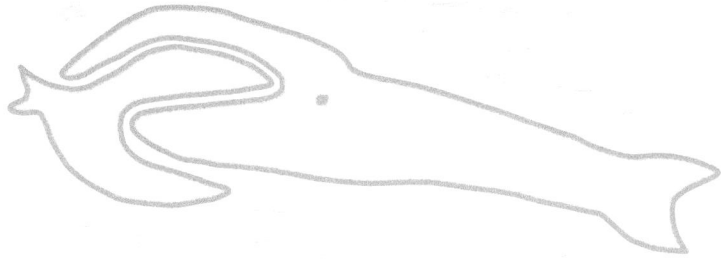

Fig. 2.1 Two sea monsters engaged in mortal combat, or a child's view of his home town?

© The Author(s), under exclusive license to Springer Nature Switzerland AG 2024   **11**
G. Notarbartolo di Sciara, *Sailing Across a Wounded Sea*,
https://doi.org/10.1007/978-3-031-54597-9_2

Two sea monsters appeared to stand out in the drawing, locked together in mortal combat; the bigger one to the right, likely a whale, was about to prevail. This was, at least, how my teacher had interpreted my work. I was nonplussed by such a lack of intuition.

My intention had been much less dramatic and more personal: I had been drawing a map of Venice, my city. My sketch was a rather naïve rendition of the extraordinary print of a bird's-eye view of Venice that was hanging with other old prints from the wall of a long, dimly lit corridor at home, which had bewitched me since the first time I had set my eyes on it. The black dot that the Maestra had interpreted as the whale's eye was meant to indicate the most cardinal of my locations: birthplace and home.

That print on the corridor wall was a famous xylograph made in the year 1500 by Jacopo de' Barbari.[1] Unlike my drawing, de' Barbari's is a work of art and, at the same time, an invaluable source of historical documentation, crafted in exact detail with all of Venice's island blocks, *rii* (canals), bridges, *campi* and *campielli* (squares large and small), *calli, salizade, rughe, fondamente* (various types of streets); the opulent palaces with their gardens drawn near the many churches with their bell towers, and the warehouses, monasteries, nunneries, hospitals, cemeteries, taverns, and everything else that makes up that city. With few notable exceptions, such as the bridge of Rialto that was still made of wood in that epoch, all the details in de' Barbari's work are still current, attesting to Venice's uniquely conservative urban make-up. The impressive precision and density of information that de' Barbari managed to cram into his work—a view from the sky as if de' Barbari could have taken to the air like a seagull—is a faithful rendition of Venice itself: constricted within a puny 5 km from end to end, the city is a concentrate of exquisite detail.

Venice is special for me not only because I was born there; it is an unearthly place where land and water are inextricably intertwined. Most medieval cities in Italy are surrounded by high, sombre walls erected for their protection in times when enemies came from all sides. Surrounding their place with water was the Venetians' unorthodox solution to national security; besides, the absence of walls suffuses the urban landscape with a most unusual light and airiness, joyful for the claustrophobics amongst us, which also frees the city from the toxic embrace of shabby outskirts, a common curse for municipalities the world over.

Water is what makes Venice unique; her domination of the seas was what made her powerful; the consequent affluence was what made her so elegant.

---

[1] Jacopo de' Barbari's story and that of his most famous artwork can be found here: https://tinyurl.com/bdu2fzmt.

Sadly, water is also what makes the city so vulnerable today, with the unfolding tragedy of human-caused climate breakdown—toxically combined with the ineptitude of the current powers in charge—that is likely to seal her fate through the unforgiving rise of sea level.[2]

Rushing in from the Adriatic Sea twice a day under the pull of the tide, seawater pours like a river into the Lagoon where Venice is nested through three narrow openings—Lido, Malamocco, and Chioggia—and penetrates the multitude of canals that make up the city's urban tissue like capillaries. Again, twice a day, water flows back into the Adriatic with the ebbing tide, conveniently carrying away the smelly organic signature of the Venetians. Despite its modest range, which is the norm in the Mediterranean, the tide functions well as the lungs and kidneys of Venice. Only on some winter days, when strong Sirocco winds push the seas up the Adriatic cul-de-sac, abnormal amounts of water enter the Lagoon and the city gets flooded, famously or infamously, by the *acqua alta*. Otherwise, it is a quiet, ordinary, moon-induced dialogue between sea and Lagoon dating back to the dawn of time, aeons before the founders, running from the Barbarian hordes from the east, decided to build their stilted abodes in the fifth century of the Common Era.

As Venice accrued block after block during her 1600 years of history to transform the original miserable hamlet of Rialto into today's magnificence, the interweaving of houses, of canals surrounding them, and of pedestrian walkways to enable all manners of urban connectivity took shape seemingly haphazardly in a balance of spaces, geometries, materials and ornaments that is miraculously exempt from bad taste: a feat that no planned design could ever have produced. Being a water city, Venice perfectly fulfils the challenging requirement that every point on the map be easily reached by foot and by boat. This dual mode of mobility can be an uncanny experience for the uninitiated. You can stroll from one side to the other of Venice, making yourself a mental image of your progress, and then repeat the passage between the same points aboard a gondola over the water, and your mind struggles to realise that you have replicated the same journey; you get the sensation of having travelled across different spaces—because not only the tortuous itineraries but also the viewpoints from land are so different from those at water level. So, it

---

[2] As climate disruption and global warming proceeds unchecked at this moment in time, I cannot see how Venice can be prevented from ending up underwater some day in a not-too-distant future. This will be long after my death; still, I cannot view this eventuality with passive acceptance. I am dreaming of a solution in which fixed bulwarks will be erected at the three openings of the Lagoon – San Nicolò, Malamocco and Chioggia – to keep rising sea level from flooding the city, with water circulation in the Lagoon assured by huge pumps constantly forcing out sea water at the same rate at which it is being let in from the higher sea level. Could this be feasible? I don't know. But I do not see many other alternatives to leaving the city to sink to her demise.

happens that when you reach a destination by boat that you had reached before by foot, you feel slightly bewildered, like when your perception has been fooled with a good trick of cards.

Back to my primary school days, I have now become amenable to viewing Mrs Angela's misinterpretation of my drawing of Venice with greater benevolence than I did at the time. It is indeed a serendipitous coincidence that the system of islands making up Venice had come to take the shape of two ocean giants, the rims of their interlocked jaws described by the sinusoid of the Canal Grande. The likeness of Venice's layout to the shapes of marine creatures was not a particularly original thought anyway, for others in the past had seen the form of a dolphin, or a fish in the city's plan. Venice's connection with the sea was to take many dimensions, one of them being how she appears from a gull's eye.

To see the semblances of animals wherever I could, was something I did very often, having been one of those children who placed everything natural at the very top of their passions and interests. This behavioural trait leads me to a consideration that goes well beyond, and hopefully justifies, the narrow scope of these autobiographical notes: because there is nothing special about a kid being receptive to the pull of nature and feeling mesmerised when watching the lives of wild animals. The tendency to seek connections with nature and other forms of life is innate in humans, according to Edward O. Wilson, which he described in his "biophilia hypothesis".[3] It was Erich Fromm the first to use the term to describe the psychological tendency of humans to be attracted to all that is alive and vital; E.O. Wilson further developed the concept suggesting that biophilia involves "the connections that human beings subconsciously seek with the rest of life". That children are so commonly attracted by animals is everyone's experience.

In my case, I could not avoid going even one step further, with the intuition that all the non-human animals I was interacting with stood at the same hierarchical level as me. That is, not as human (me, in my case) vs. animal, but as animal (in my case, human) vs. another animal, e.g. a cat, a dog, or anything else. This intuition was derived from the inescapable observation that these other animals had personality traits just like my own, and were capable of sentiments and emotions (such as excitement, sadness, joy, fear, and hopefulness). To the point that I saw no valid reasons for not considering them as persons just like me. This way of viewing non-human animals—which I think should come naturally to anyone having had anything to do with non-human

---

[3] Wilson E.O. 1984. Biophilia. Harvard University Press. 180 p.

animals—had obviously significant implications in my relationship with the natural world.

Unfortunately, however, something goes amiss in such a relationship as children grow. Something that I find sad and difficult to accept, noting that so few children retain such a fascination for life and nature when they grow up. Whatever the reasons, humans' progressive detachment from the natural world as they become adults, at times even resulting in biophobia[4]—urbanites wishing to raise the drawbridge, lower the portcullis, and retreat into their protective nature-proof cocoon—is all so detrimental to our relationship with the planet.

The sensation of being immersed and integrated within a landscape still unblemished by the human hand can be a source of pleasure that borders on sensuality, and works for me as a conduit to happiness. Meadows and forests, oceans and mountains, rivers and lakes have a powerful effect on my mood because they all exude a sense of being alive, and infuse a physical meaning into my own sense of being alive. It feels like I am craving a daily dose of nature like a drug, which has never abated during my existence. Alas, aside from the symbolic coincidence of being shaped like two wrestling leviathans, Venice had not been generous in terms of allowing me to be exposed to wildlife during childhood. Today, the Venice Lagoon hosts a much greater wildlife richness—birds, for example—than it did seventy years ago. At a very early stage of my existence, I felt that my starvation for the natural world could no longer be remediated by the company of city pigeons, stray cats, and the occasional furtive rat.

This all changed the day I found out that what I was searching for was hiding under my nose. One afternoon, my mother took me with her to the Venice Golf Club. She was going to play with her friends, and I was to be around—a good opportunity for being outdoors in the beautiful estate of the golf course, situated at the south end of the Lido in a locality called Alberoni, very close to the Malamocco opening of the Lagoon to the Adriatic Sea. Whilst liking the idea of spending an afternoon in a nice natural setting, I knew I was also in for a significant boredom session. I could not comprehend this passion for chasing around a ball with a stick across the fields, and besides, as a 10-year-old, I was not going to play golf anyway, and the conversation of the grown-ups could not have been less interesting. So, as soon as I could, I excused myself from the company of Mother and her party and took off in exploration.

---

[4] Harwitz E. 2023. Beware creeping biophobia. Hakai Magazine http://bit.ly/3yXmSY2.

Access to the course's premises was through a bridge across a quiet canal surrounding the property. Nobody was around, and the water of the canal, unusually transparent at that moment because of the incoming tide from the near opening to the sea, was beckoning with promise and attracted me like a mesmerist. Lying on my tummy on a skiff someone had moored on the side of the canal near the bridge, my nose glued to the still surface, an extraordinary underwater theatre pulsating with life revealed itself to my stunned eyes, perfectly visible as through the glass of an aquarium. Dominating the scene was a diverse, incredibly elegant assemblage of sea anemones covering the rocks scattered over the shallow bottom, slowly swaying back and forth their pastel-coloured tentacles (Fig. 2.2).

The main actors in the play were a handful of hermit crabs sporting an assortment of shells, scuttling with haste across the stage, which was aspersed with mounds and holes made by god-knows-which mysterious burrowing critters. There on the left, half-disguised under a canopy of bright green lettuce algae, a shore crab was lurking threateningly. Even a brittle star, something I had never set my eyes upon before, was crawling along in a hurry, expertly pushing forward through the movement of its serpentine arms. This

Fig. 2.2 Taken from an old book I found in my home library (*Figuier L. 1882. Molluschi e zoofiti. Fratelli Treves Editori, Milano, p. 369*), this engraving illustrating a variety of sea anemones replicates quite closely my memory of the spectacle appeared in front of my eyes during my childhood across the clear waters of a canal in the Venice Lagoon

company of characters was putting up a hell of a show in that fantastic under-water micro-theatre to the benefit of their one-kid public, and the kid was ecstatic.

Seemingly unimportant episodes, serendipitously occurred, can have the power to change the course of one's life. Whether this distant epiphany with the minuscule underwater show changed mine, I do not know. More impor-tantly, though, that episode taught me that the natural world we live in can be a source of hidden fascination, to be enjoyed if one has the luck of stumbling on it and the ability to see it. Not all of these gems need to be large and char-ismatic like the whales, the dolphins, or the sharks I had the opportunity to admire later in my life. Some are small and hidden, like the sea anemones, hermit crabs and brittle stars in a canal of the Venice Lagoon. Even more importantly, it is not just the characters that matter but the play they are in. It is not only the single species per se, or even their sum, that create exception-ality—but the system they are part of, which they concur to keep alive and in good health with their existence and constant toiling. Put more technically, it is ecosystem functioning rather than just ecosystem structure that com-mands awe.

In the several decades succeeding my childhood in Venice, I was fortunate to acquire knowledge of the natural life in the Mediterranean Sea and beyond, that replicated that initial experience of discovery, albeit scaled up enormously. Back in 1976, having terminated my first curriculum of studies, I had the chance to meet in Venice the extraordinary oceanographer Walter Munk[5] and his wife Judy, thanks to the introduction by a shared acquaintance. In that momentous, life-changing event, Walter invited me for a brief visit to his home in La Jolla, California, a renowned temple of marine science, where he would support my involvement in marine biology and introduce me to a few key players in the field. The "brief" visit ended with me staying with the Munks for nine years, generating a lasting, immensely enriching, quasi-father-son relationship with my mentor. Aside from the completion of my academic curriculum, those 9 years saw a sequence of events which created in me a last-ing connection with the world of science and conservation of marine mega-fauna, first with humpback whales in Hawaii, then with Bryde's whales and manta rays in Venezuela, and finally with devil rays in the Gulf of California.

Back in Italy in 1985, I was surprised to discover that my country was unwilling, or at least unprepared, to embrace a "prodigal son" with a profes-sional profile like mine. It was only through the almost reckless initiative of

---

[5] Walter H. Munk was an outstanding 20th-century oceanographer. His biography is briefly described here (including a mention of the devil ray I dedicated to him): https://tinyurl.com/25p658jc.

inventing a marine research organisation out of thin air—the Tethys Research Institute—that I managed to create the conditions for cultivating marine science in a society that had become extraordinarily science-unfriendly, and had forgotten the country's past glories in the field of the advancement of scientific knowledge.

The professional journey briefly described above has been, for me, a great privilege, only hindered by two disturbing sentiments: a sense of precariousness and one of solitude. Precariousness because of the nagging, angst-provoking awareness of the impending danger that humankind is subjecting to the objects of my discoveries. Solitude derived from considering that if the knowledge I was gaining could have been more widely shared with my fellow humans, including the person in the street and not just the handful of converted within my narrow working entourage, perhaps more would embrace an attitude of greater respect and appreciation for the Mediterranean environment and its inhabitants. "One of the penalties of an ecological education is that one lives alone in a world of wounds", wrote Aldo Leopold shortly before I had my lagoonal epiphany,[6] words that I have recalled more often than I wished.

Something had to be done. I wanted to create conditions amenable to opening my heart to my human fellows to get them to share my concerns, eliminate this sense of loneliness, and stimulate the growth of a greater ecological conscience that may one day decrease human-caused pressure on the Mediterranean environment. Thus was hatched the idea of embarking on a journey across the Mediterranean to visit and celebrate some of these non-human marine characters I was so lucky to have encountered in my life; and introduce them to readers who might be amenable to partake in the discovery and maybe even to let themselves become a bit transformed by such discovery.

With this intent in mind, one day in early June in the early twenties of the second millennium, I ventured out of the protective womb of the Venice Lagoon aboard a sailing vessel named *Pontoporia*. I set sail to taste the salt of the open sea with a cargo of expectations and anticipation. My plans for this 4600-mile journey were ambitious. I would first descend to the southern end of the long and narrow Adriatic Sea, an elongated branch of the greater Mediterranean sandwiched between the Italian and Balkan peninsulas. Once out of the Adriatic, I was to hug the Greek coastline along the eastern Ionian Sea, and from there, across the Corinth Gulf and Canal, I would penetrate the southern Aegean Sea, to be navigated across its longitude from west to east. Next was going to be a foray into the Mediterranean far east, the Levantine

---

[6] Leopold A. 1949. A Sand County Almanac. Oxford University Press.

Sea, first along the south coast of the Anatolian Peninsula, then south of Cyprus, the Holy Land, and finally the Nile Delta, bound to Alexandria. Having reached the sea's easternmost extreme, I would then set my course to the west to southern Crete, and from there, I would make the long passage to the Strait of Sicily with a stop in Lampedusa. A quick visit to the Egadi islands at the west tip of Sicily would follow, with another brief stop in the south of Sardinia, and then another longer passage to the gates of the Mediterranean at Gibraltar, possibly dipping the nose in the cool Atlantic waters. The last leg of the journey was going to be from Gibraltar across the Alborán Sea, south of the Balearic Islands, and finally across the Gulf of Lion to Sanremo, in the Ligurian Sea, with a grand finale in the "Pelagos Sanctuary for Mediterranean Marine Mammals".

3

Adriatic Sea: From Venice to Otranto

Illustration by Massimo Demma

© The Author(s), under exclusive license to Springer Nature Switzerland AG 2024
G. Notarbartolo di Sciara, *Sailing Across a Wounded Sea*,
https://doi.org/10.1007/978-3-031-54597-9_3

Day 1: Venice, Italy

At the strokes of the Marangona—the enormous, bass-pitched bell tolling from San Marco's campanile at eight in the morning to mark the beginning of the workday at Venice's Arsenal—*Pontoporia* had already left the San Marco Basin and was negotiating the narrows between Sant'Andrea's Fort and San Nicoletto at the northern tip of the Lido. I was grateful for the kick in the pants received from an ebbing tide. Soon, the colour of the water seen beyond the gunwale was turning from the lagoonal pea soup tint to turquoise blue, with the greenish hue progressively decreasing to indicate my transition to marine surroundings. As I proceeded to the southeast by hugging Lido's northernmost breakwater, I was soon out into the Adriatic Sea. Venice, her bell towers and all her saints were now in my wake, slowly fading away in the morning haze as the boat ventured out into the open waters.

It was around the beginning of summer, and that windless day was going to be hot and hazy. Visibility was poor, and the land on the other side of the Adriatic, where I was directed, was not in sight. *Pontoporia* was proceeding well under power; I was pushing her at almost ten knots, well beyond the seven knots that are her optimal cruising speed, because I was keen on arriving at our first day's destination before dark.

In front of *Pontoporia*'s bow, the vast marine expanse was beckoning with the promise of encounters with the various characters composing the Mediterranean "charismatic megafauna": a category including the large animal species, often faring in a precarious state of conservation, that have symbolic value and widespread popular appeal and that can support environmental protection goals because the public likes them. Contrary to widespread prejudice, the Mediterranean Sea hosts a rich and diverse charismatic megafauna, including marine mammals (i.e. whales, dolphins, and the monk seal), marine birds, marine turtles, sharks, and rays. Much of what is known about these animals, beyond the mere notion of their existence—e.g. their habitat, behaviour, ecology, population sizes, and main threats to their survival - was obtained by research efforts that happened only during the past two to three decades. Admittedly, the Mediterranean is not as dense with these megavertebrates as many other portions of the world ocean. Besides, large predators, the main components of the category, are bound to be naturally rare everywhere they occur, and not just here in the Mediterranean, because ecosystem balance never leaves much room for large predators at the top of the ecological pyramid.

Furthermore, large predators are more frequent in regions where the waters are highly productive, which is not the case in the Mediterranean except for some specific and limited locations. Add to the mix the mindless environmental abuse exerted by humans on this region and its inhabitants for at least the past hundred 50 years, with actions particularly detrimental to large, slow-growing, and slow-reproducing species such as the whales, the dolphins, the seals, and the sharks. It becomes clear why these waters are even emptier today of such large critters than they would have been under more favourable circumstances. And yet, these animals are still here, hiding in plain sight of their human neighbours.

Dedicating excessive attention to charismatic megafauna has raised more than one eyebrow from most purists amongst conservation scientists, under the argument that all species deserve consideration regardless of their level of flamboyance. Point is well taken, of course. This position, however, must not lead to the concrete risk of dismissing concern for the conservation of some species simply because they are charismatic. It is a sad fact that in the past, particularly in the Mediterranean, marine megafauna has been egregiously neglected, not just by society in general but by most zoologists as well. Until well into the second half of the twentieth century, scientists were in the thickest of fogs, even simply about the notion of which species of whales and dolphins occur in the Mediterranean. Many otherwise well-respected zoologists, for example as late as the 1970s, were blissfully unaware of the presence in the region of a resident population of fin whales numbering in the thousands.

The Adriatic's northernmost portion, which is as far north as one can be anywhere in the Mediterranean, contains the region's largest shelf area and is so incredibly shallow—about 20 m deep on average—that you could be forgiven for not being sure of having come out of the Lagoon and being already out into the open sea. Except that it was not quite like navigating in a lagoon because—despite the lack of wind—the surface of the water had now taken that almost imperceptible undulation that is a prerogative of the open sea and can hardly be perceived if it were not for the gentle moving reflection of the sun on the glassy surface. The undulation exists because the ocean is never still, even on the calmest days, and its slow breathing, generated perhaps by some distant storm, is always there to warn the sailor that its sleepy mood will not last forever.

With the first stop of our journey, I intended to visit a healthy community of common bottlenose dolphins,[1] denizens of a small body of water called

---

[1] Bottlenose dolphins, *Tursiops truncatus*, are called "common bottlenose dolphins" in English to distinguish them from the tropical Indo-Pacific bottlenose dolphins, *Tursiops aduncus*, which occur in the Red

Kvarnerić, just north of the Dalmatian Archipelago in Croatia, comprised within the islands of Cres, Lošinj, Krk, and Rab. To highlight the importance of this part of the Adriatic for bottlenose dolphins, in 2016, we identified here a "Northern Adriatic Important Marine Mammal Area", in short, an IMMA. IMMAs are portions of habitat important for marine mammal species, identified through the application of scientific criteria on the basis of expert opinion and results from research efforts. They are not marine protected areas because they are delimited by scientists, not by legal authorities, but they are designed to attract the attention of authorities to the need to protect these areas. Like for all IMMAs, information about the criteria that were applied and the characteristics of this area can be found on the online atlas posted on the website of the IUCN Marine Mammal Protected Areas Task Force.[2]

To reach my destination, I had to sail around Cape Kamenjak, the southernmost tip of the Istrian Peninsula, a 75-mile leg from Venice. As the day progressed, towards noon, the haze around the horizon had lifted, and the relief of the rocky Croatian coast had come clearly in sight, becoming increasingly distinct as I was approaching the land on the opposite side of the Adriatic.

This first day of navigation was not brightened by any wildlife encounter worthy of note, except for a few scattered loggerhead turtles[3] along the way. They were bobbing at the surface with their rounded heads well raised above the water to watch my approach in mild alarm, all of them diving in haste when *Pontoporia* passed at a distance they deemed too close for their comfort. Loggerhead turtles are the most common marine turtles in the Mediterranean Sea. Once rare and considered under threat, their numbers have recently increased, bringing the population out of the danger zone, primarily due to

---

Sea, Persian Gulf, Indian Ocean, and West Pacific Ocean, but not in the Mediterranean Sea. In these pages "common bottlenose dolphins" are simply referred to as "bottlenose dolphins".

[2] Northern Adriatic Important Marine Mammal Area: https://tinyurl.com/kus7s5uf.

[3] The Mediterranean subpopulation of the loggerhead turtle, *Caretta caretta*, is now assessed as "Least Concern" in IUCN's Red List of Threatened Species: https://www.iucnredlist.org/species/83644804/83646294.

the protection of their nesting sites—mostly along eastern Mediterranean beaches, thanks to the efforts of a galaxy of dedicated non-governmental organisations (NGOs). The northern Adriatic is an important foraging ground for these omnivorous reptiles, and they congregate here in large numbers to feed on just about anything that lives on the bottom or swims in the water column—and is clumsy enough to lend itself to be snatched by this equally clumsy predator or scavenger, depending on the circumstances.

Aside from the turtles and the occasional gull hurriedly flying overhead, there was not much else to be seen on that day regarding wildlife. But I was not expecting otherwise. In addition to the first-day excitement of being finally out at sea in the sole company of *Pontoporia*, with all the expectations of wonders and adventures ahead, I had noted during many months spent alone on vessels in the different circumstances of my life that being at sea is never dull. There is something about pushing a boat forward that becomes part of one's baseline physiological activity, like walking or breathing. I knew that from the days I had spent in my younger years rowing alone on the stern of a gondola in the Venice Lagoon, when the forward motion of the boat resulted from leaning on the single long oar with the weight of my body, without the exertion of significant effort. I could do that for hours on end without feeling tired because of the perfect design of gondolas and their way of being propelled by the single, asymmetric oar, honed across centuries of use as if it had evolved in a living organism. Here aboard *Pontoporia* it was not me but the wind, or the engine, causing the boat to advance. But I had to steer her nevertheless, and it was a very similar feeling of being one with the boat, with the task of keeping the course steady delegated to some lesser, mechanistic layer of the brain.

As I doubled Istria's southward projecting headland around mid-afternoon, I entered the Kvarner Gulf, framed between Croatia's mainland and the island of Cres, a mere 20 miles away. However, I was not going to be able to reach my destination that day, on the east coast of the island of Lošinj. The fastest way to enter the Kvarnerić from where I was involved negotiating the narrows between Cres and Lošinj in a location called Osor. Osor is placed on an isthmus connecting the two islands. A narrow channel was dug in Roman times across the isthmus to facilitate maritime travel and transportation in the region. Today, the two islands are connected by a road crossing that narrows over a swing bridge that opens twice daily to allow for the north-south passage of marine traffic. It was going to be late for the afternoon opening. Still, the weather was fine, and there were plenty of hospitable indentations along the west coast of Cres where I could drop anchor and spend the night, waiting for the Osor bridge to open the following morning. The bay of Ustine, in particular, perfectly suited my needs: three miles north of Osor, it stood a mere 20 min of sailing from the bridge. Slowly coming into the pretty bay just as the sun was setting, surrounded by lush Mediterranean vegetation, I was welcomed by a chorus of cicadas tiredly prepared to range away their musical instruments for the night.

I had left Venice only hours before, and yet it seemed to me that I had landed on a different planet. Here, the walls were made of trees instead of stone, the water was blue and clear instead of green and soupy, birdsong had replaced the sound of church bells, and the air was full of the balmy fragrance of the Mediterranean maquis.

Day 2. Ustine, Croatia

Osor, a place of importance in ancient times, is today a small town of no more than 60 souls. Its historical pedigree is revealed not only by its position as a crossroads both for land and sea transportation, which gives it a sense of consequence by itself, but also by the many layers of ruins that surround the village, from prehistoric to Greek, Roman, Byzantine, and finally, Venetian, dominated by a sixteenth century's gothic cathedral dedicated to Bishop Gaudentius. Almost as importantly, Osor has a cafe where I could enjoy a decent cappuccino after having sailed across the strait and temporarily docked *Pontoporia* off the southern pier. Although *Pontoporia*'s galley was equipped with the Moka coffeemaker that no Italian would ever leave port without, the attraction of a decently made cappuccino is never to be resisted when available.

The leg from Osor to Rovenska, a small harbour on the east coast of Lošinj, was only slightly longer than the one from Ustine to Osor. The first portion involved navigating across a very shallow bay. With *Pontoporia* being a sailing boat with a respectable draught, I had to carefully follow a channel dug along the middle, and marked with a series of buoys, to avoid getting stuck in the muddy shoals. But soon, I was out into the deeper blue waters of the Kvarnerić, on that day scattered with white sails in the picturesque surroundings.

The Kvarnerić still provides an excellent habitat to common bottlenose dolphins. I had a long story of affiliation with the place, where I first came in 1992 to support Giovanni Bearzi, at the time a student of mine, to establish a research base dedicated to the study of these dolphins and their habitat. This activity was part of the operations of the Tethys Research Institute, which I was directing in those years. Giovanni had been visiting the place since 1987 with that purpose, and it eventually became evident that his project deserved the establishment of a more permanent base and of a formal research programme, the "Adriatic Dolphin Project", operating in what still was, at the time, the Socialist Federal Republic of Yugoslavia.

In those years, not much was known about the ecology and conservation status of bottlenose dolphins, neither in the Adriatic nor in the wider Mediterranean for that matter, and the Kvarnerić provided an excellent set-up for facilitating scientific discovery. The place was reasonably accessible by road from northern Italy, where Tethys is based, and the dolphins appeared to be abundant and easily found in the Kvarnerić's sheltered waters. During the years, Giovanni and other colleagues who joined him at the Adriatic Dolphin Project were able to cover lots of new ground by following the dolphins with small inflatables for hours, day after day, and year after year, cataloguing more than one hundred different dolphins identified through high-quality photographs based on individually distinctive, permanent body marks, such as nicks in their dorsal fins or specific traits of their colouration.

Research efforts in the Kvarnerić revealed, amongst other things, that the dolphins' range was significantly greater than the study area (about 800 km²), that the dolphins remained in the area consistently across years (in scientific jargon, they had a high level of "site-fidelity"), that dolphin group size averaged seven animals, but the groups of females with their young were larger than the groups of adult males which lived separately, and that dolphins spent a greater portion of their time budget than their conspecifics from other parts of the world hunting for fish, indicating a likely condition of prey depletion in the area.[4]

Meanwhile, during the first half of the 1990s, all hell had broken loose in Yugoslavia, with parts of the Federation, Croatia included, fighting for their independence from the central state in a bloody internecine war. The study area of the Adriatic Dolphin Project was never directly affected by the hostilities, so we decided to continue with the research plans there despite the obvious adversities and uncertainties deriving from operating in a country at war. Meanwhile, over the years, the project was joined by increasing numbers of local young and passionate participants, including Draško Holcer. Having made his first steps from within the ranks of the Adriatic Dolphin Project, Draško is recognised today as Croatia's foremost expert in the field of marine mammal ecology and conservation and works as the senior curator of the marine zoology collection of Zagreb's Croatian Natural History Museum.

---

[4] Bearzi G., Notarbartolo di Sciara G., Politi E. 1997. Social ecology of bottlenose dolphins in the Kvarneric (northern Adriatic Sea). Marine Mammal Science 13(4):650-668 and Bearzi G., Politi E., Notarbartolo di Sciara G. 1999. Diurnal behavior of free-ranging bottlenose dolphins in the Kvarneric (northern Adriatic Sea). Marine Mammal Science 15(4):1065–1097.

Eventually, having discovered another location holding high promise for marine mammal research in Greece—which I was set to visit in the coming weeks—and finding it challenging to engage in both efforts at the same time, Tethys decided in 1999 to pass the baton of the Adriatic Dolphin Project to our Croatian friends, under the umbrella of the Blue World Institute, a research organisation created by Draško and his colleagues.

When I entered the small harbour of Rovenska slightly before noon, I was greeted by Draško on the pier as he helped me tie the lines to the dock. Our plans for the day were to take a short trip around the Kvarnerić in search of the dolphins with the Blue World Institute's inflatable outboard, leaving *Pontoporia* to rest in Rovenska.

As soon as we were back at sea, the old excitement of stalking wildlife sent a shiver down my back. A very benign hunt indeed, in which no malicious effect was imposed on the quarries, but with all the traditional elements of hunting save the violent conclusion: scanning the horizon in search for the telltale disturbance of the water surface, the surge in concentration when something which might have been dolphins is spotted in the distance, the confirmation that dolphins are indeed there, the cautious approach at slow speed to avoid spooking the animals with unwanted noise and avoidable intrusiveness. All of the above, of course, was facilitated by the fact that the dolphins in the area knew all about the research team—who they were and what their behaviour might be like—and had no reason to do anything different than what they would have normally done, had the researchers never existed. This observation leads me to suspect that the dolphins have an expertise of humans much more sophisticated than the expertise of dolphins we humans have accumulated during years of observations (Fig. 3.1).

By following the dolphins day after day and year after year, researchers seize what opportunities would present themselves for taking good-quality photographs of their distinctive dorsal fins to recognise the different individuals later and reconstruct who they were associating with. Very soon, in our observations of the dolphins in the Kvarnerić, we discovered how dynamic their individual relationships were. Some dolphins liked to stay always together,

Fig. 3.1 A young bottlenose dolphin, porpoising out of the water, shows the short and rounded rostrum typical of its species, and the grey colouration of its flank devoid of any specific distinctive pattern (Illustration by Massimo Demma)

mothers with their young being the clearest example. Others would associate only for hours or even minutes. Following a group of dolphins for a few hours, we noted that the membership within any single group changed constantly, and the only way to monitor group dynamics was by recognising them individually.

Such recognition effort was based on the quite effective method of photo identification. This technique, used to study small populations of individually identifiable wild mammals, including dolphins and whales, is very powerful despite its low-tech appearance. Not only is it possible to understand many aspects of social structure and dynamics when observing repeatedly known individual animals. By comparing the numbers of individuals one has "captured"—albeit only virtually, on photo—at different times, it is possible to derive the size of the population those individuals belong to; and handy software exists today to aid in these calculations, including assessing the level of uncertainty involved in the results.[5]

Studies carried out the world over, based on the recognition of individuals identifiable via good quality photographs taken at different times, had revealed that bottlenose dolphin societies are made of larger units composed of females—several generations of them, grandmothers, mothers, with their calves of both sexes—and smaller units made of two to three adult males. Tethys' research in the Kvarnerić demonstrated that those patterns described from observations made in various marine regions also applied here: clearly, a species-wide trait. However, these smaller units would then often merge into larger units into a highly dynamic process, the so-called "fission/fusion"

---

pattern, depending on whether the dominant behaviour of the moment was travelling, feeding, socialising, mating, or resting.

A strange yellowish haze was hanging over the horizon on that windless day. The sky had turned from blue to whitish. The weather forecast had warned about a developing state of instability, and I could tell that a storm was in the air somewhere. But the sea surface was calm like oil, which makes all the difference in the world when searching for marine animals or observing them once found. Still, it takes a trained eye to detect the minuscule, dark, pointy tips of the animals' dorsal fins breaking the surface in the distance as they progress: an almost imperceptible disturbance close to the horizon's linear monotony but a clear, unambiguous sign of the dolphins' presence.

And in fact, we were lucky because it didn't take long to find the dolphins on that day. So it was that on the second day of my cruise, I finally managed to make contact with bottlenose dolphins, one of the main representatives of the Mediterranean marine megafauna: the first in a long collection of encounters I was to make in the following weeks.

That day, we sighted a group of nine dolphins, and Draško knew them well: a unit of seven females with one calf, joined later by two adult males. One of them, whom we had called *Freccia Bianca,* was an old timer first sighted in the early days of the Adriatic Dolphin Project, still happily swimming around three decades later. Long-term research projects have revealed that bottlenose dolphins can live sixty years or more in the wild.

The behaviour of the dolphins did not seem to be perturbed by our approach and gave us the confidence that we were not affecting their well-being in any significant way. As we followed them at a respectful distance to observe their behaviour and association patterns, the dolphins were progressing in a zig-zag course with frequent dives, probably searching for prey. Quite often, they were surfacing very close to our boat, something that they would have been able to avoid had they wished to stay away from us. Maintaining close contact with a group of wild dolphins is, for me, always a source of excitement, no matter how many times I have found myself in that condition, because every encounter is full of novelty as if it were the first. However, as it often occurs when encountering wildlife, you get the sense of regret that the glimpses of the animals are too fleeting and that you wish to have opportunities to see them better and for a longer time than they allow you to. Dolphins can swim fast and can stay underwater and out of sight for most of their time, if they so wish. When they surface near you, that is normally for a handful of seconds, and when they are gone, they leave you craving for more. Even when they approach to ride the bow wave of the boat you are sitting on, and you have that very special chance of watching them for many minutes at close quarters as they jostle amongst each other to occupy the prime position in front of the vessel, you get the sense that they exist in a world which is separate from yours by an invisible barrier and that real contact is out of your reach. They do their things and interact with their companions but not with you, and it is easy to feel somewhat excluded from their play. Dolphins, particularly the bottlenose kind, are also very vocal. They create their soundscape around them and their group, made of two main components: a variety of utterances—whistles, creaks, burst pulses—that they use to communicate with their schoolmates, and very high frequency "clicks" which are a way they have to inspect their environment acoustically, similar to bats, called echolocation. These noises are barely audible to the human ear if the observer is above the surface, which is why I had fitted a hydrophone to the hull of *Pontoporia*.

An essential ingredient in the excitement of participating in a project in which a group of animals is followed consistently across a period of several decades derives from the fact that as you progress in time and knowledge, the animals mutate in your perception from the abstract category they belong to—"bottlenose dolphin" in this case—into a community of individual beings, each with its defining physical properties and personality. This brings the relationship between the investigator and the investigated onto a more equal and far more meaningful level, and in fact, during some of these sessions, the concept of who investigates who becomes somewhat muddled.

As dolphins go, bottlenose dolphins are quite an interesting lot. The species occurs in all oceans, including many semi-enclosed seas such as the Gulf of Mexico, the Caribbean, the Gulf of California, the Red Sea, the Persian Gulf, the Black Sea, and, of course, the Mediterranean. The bottlenose dolphin is one of the cetaceans with the widest circumglobal distribution, and within a broad latitudinal range. For instance, in the Atlantic Ocean, bottlenose dolphins regularly range from north of the British Isles to southern Patagonia, typically occurring in tropical and temperate inshore and coastal waters. Being adaptable to various environmental conditions, including those significantly made worse by human negligence, is one of their trademarks. To their misfortune, their ease about dwelling in shallow waters has made them the optimal choice as captive subjects. This explains why bottlenose dolphins are the most common species hosted in dolphinaria.

A recent survey estimated that a minimum of 60,000 bottlenose dolphins live in the Mediterranean Sea, where they can be primarily found over the shallower portion of the basin down to a depth of approximately 200 m. The most recent population estimate in the Adriatic Sea is about 10,000 of these dolphins.[6] Given that common dolphins, the only other cetacean species known to have regularly inhabited the Adriatic, have been extirpated from this sea sometimes during the twentieth century and appear here only on fleeting occasions, bottlenose dolphins are today the only cetacean regularly found in the region. Further to the south, in the deeper waters of the lower portion of the Adriatic, the diversity of cetacean species increases significantly, with bottlenose dolphins sharing their range not only with common, striped, and Risso's dolphins but also with Cuvier's beaked whales and the occasional sperm and fin whales.

Given the intensity of human activities in the waters of Lošinj, including fishing and tourist boat traffic, the idea of creating a protected area for dolphins in the waters of Lošinj had been floated many years ago by Tethys based on Giovanni's early research, which demonstrated the importance of the area for these animals. Later on, the effort was further spearheaded by the Blue World Institute. Quite counterintuitively, that idea encountered local opposition for fear that the protective regulations might have impacted marine tourism, locally a key economic activity. However, that obstacle was later overcome after Croatia had become part of the European Union, and a Natura 2000 site for bottlenose dolphins was declared in the Kvarnerić, coinciding with the initial Cres-Lošinj protected area project. The area was also flagged as part of the Northern Adriatic IMMA, identified by the Marine Mammal Protected Areas Task Force of the International Union for the Conservation of Nature (IUCN).[7]

Draško did not seem too pessimistic about the conservation outlook for dolphins in the Kvarnerić. "There seems to be a reasonably good balance in the dolphins' relationship with the fishing activities here", said Draško, "although there are some signs of food depletion caused by fisheries".

"I agree that fisheries may not be posing a special risk to dolphin survival here", I said, "but what about disturbance to the dolphins by pleasure

---

[6] Natoli A., Genov T., Kerem D., Gonzalvo J., Holcer D., Labach H., Marsili L., Mazzariol S., Moura A.E., Amaha Öztürk A., Pardalou A., Tonay A.M., Verborgh P., Fortuna C. Submitted. *Tursiops truncatus* (Mediterranean subpopulation). The IUCN Red List of Threatened Species.

[7] Northern Adriatic Important Marine Mammal Area: https://tinyurl.com/kus7s5uf.

boating?" As I spoke, a mother and calf dolphin pair surfaced so close to our boat that I could almost touch them.

"Private boating becomes intense in the tourist season, and the potential disturbance to the dolphins is a concern", said Draško, "which we try to assuage by increasing awareness amongst tourists and recommending the adoption of a code of conduct by boaters".

"As can be seen from the behaviour of these dolphins we are following today, it is not the presence of boats in their vicinity per se that is the problem, but the behaviour of boaters who may not be aware of which is the best conduct to have with dolphins around, and what are the things to be avoided. Additionally, our research here showed that the sheer number of boats around during the summer creates quite a noisy environment, which causes dolphins to change their vocalisation patterns.

"Compared to many other Mediterranean locations, the general assumption is that this location is reasonably favourable to bottlenose dolphin conservation", continued Draško. "However, although tourist pressure disappears after the summer months, we are cautious about the long-term effect it may have on this population".

As Draško was slowly driving back the boat in the late afternoon and delivered me before sunset to the pier of Rovenska, I considered that, after all, bottlenose dolphins are not faring too poorly here in the Kvarnerić in the general context of the status of their conspecifics elsewhere in the Mediterranean. However, what was the status of the species here, say, a 100 or 200 years ago? Ancient accounts give us good reasons to think their numbers were much higher.

As the day was coming to an end, it was time to prepare for the departure on the following morning, for the longer passage that would eventually bring me to reach the heel of the Italian peninsula at the southern boundary of the Adriatic. I had anticipated having a peaceful dinner aboard, in the cosy little harbour of Rovenska, surrounded by an amphitheatre of small waterfront houses, which reminded me of the building styles of the Venetian coastal region. I had procured from a nearby tavern a dish of palačinke, the delicious Croatian crepes, which were to be followed by the juiciest season's cherries and apricots and a slice of Asiago cheese I had brought with me from Venice. As I was getting ready to eat on the small table I had dressed out in the cockpit of *Pontoporia*, fearsome cumulonimbus clouds, blacker than black, were gathering over the Televrina, Lošinj's highest mountain, impending overhead, blocking the sunlight and suffusing the eerily silent calm with an ominous sense of foreboding. It was the proverbial calm before the storm—and quite a storm was the one that followed.

Uninvited, Neverin (a meteorological phenomenon well-known in that part of the Mediterranean, famous for its short life but also its raging violence, potentially deadly if it catches you offshore) was going to be my troublesome company for dinner. First came the blinding light of a thunderbolt striking the bell tower of the nearby church, no more than a few 100 metres away. Then was the turn of the first gust of wind, luckily from the northwest, which ensured that the small harbour of Rovenska turned out to be the safest place I could be in, but so potent and sudden to cause *Pontoporia*'s mooring lines to creak under the strain, with the prized palačinke threatening to fly away from the plate in the process. Then, a wall of rain started roaring down from the mountain, giving me just enough time to scramble inside the deckhouse, ultimately saving the palačinke from miserably drowning under the punishing downpour. But the inside quarters of *Pontoporia* were nice and dry, and with the boat safely harnessed at the dock, unfazed by the irate winds blowing from land, the feeling of protection was so palpable that—having finally wolfed down palačinke, Asiago, and apricots—I passed out in my bunk.

Day 3. Rovenska, Croatia

The new day was announced by the melodies of an inspired male blackbird venting out the claim to his territory with operatic flair. The sky was clear, and the air calm and cool; the tender leaves violently ripped off the trees by yesterday's maelstrom, scattered inside the puddles created on the quay by the night's rains, were the only signs that the Neverin had not occurred solely in my dreams.

I left the dock leisurely around mid-morning, mindful that the passage to my next destination—the town of Otranto in the southern portion of Apulia, in Italy—required approximately two full days of navigation. The forecast was excellent, with a high-pressure regime stabilising again over the central Mediterranean. I expected fair weather over the next few days, which was great given that I planned to continue navigating non-stop, day and night, until I reached Otranto. Engaging in a solo voyage in a sea busy with traffic is always a challenge when the plans include navigating day and night, for the lone sailor cannot be relieved in a shift by crew mates. During this journey, I would have to endure nocturnal navigation several times, which was made possible in my solitary condition by an increasingly helpful technology. The most significant support came from the ability to know my position at any moment and with utter precision with GPS, a device so named under its acronym that stands for "Global Positioning System". Sailors today take it for granted that they can know where they are because GPS has become ubiquitous, even on the cheapest of smartphones. Initially developed by the United States to assist in military operations, the service was made available globally for civilian use in the 1990s by President Clinton and Vice-President Al Gore. GPS is based on analysing the different times of arrival of signals from a network of overhead satellites, from which a position can be derived with the extraordinary accuracy of a few metres. I remember well the sense of marvel when trying out an early, bulky model of GPS receiver here in the Kvarnerić, and considering that the string of numbers appearing on the small screen, corresponding with the coordinates I was in, was nothing less than sorcery. But *Pontoporia*'s technological paraphernalia went much beyond GPS. The autopilot kept the boat heading steadily on a pre-set course, freeing me from constantly staying at the helm and concentrating on keeping the bow in the desired direction. My safety was significantly enhanced by a small version of the good old radar, which would sound an alarm when another vessel would come within a set radius of my boat. The radar's job was further supported by the newer Automatic Identification System (AIS), through which vessels automatically broadcast their position to land via radio signal, and from there, the information is provided to anyone having the appropriate receiver. This suite of apparatuses—part of *Pontoporia*'s safety equipment—allowed me to sail as long as necessary, including at night, and gave me the luxury of short, semi-vigilant power naps I had become used to until daylight would come again.

I sailed inside the Kvarnerić for a handful of miles hugging the southeast coast of Lošinj, green with its unspoiled cover of pine trees reaching down to the water line, until I crossed the wide passage between Lošinj and the small island of Ilovik, thereby exiting into the open Adriatic. As I was getting farther from the coast, the air was as calm as it could be on that warm day, the Neverin of the day before a distant memory with the seas flat—needless to say—like a palačinke. The plan was to navigate down the length of the Adriatic, mainly along the sea's eastern coasts, rocky and scattered with the many Dalmatian islands, and where the waters—unlike along the sandy coastline along the Italian shore to the west—are blue and clear.

As I ventured out into the open, I could not avoid feeling a sense of sorrow for the poor Adriatic, one of the world's most generous providers of seafood of the highest quality and, at the same time, one of the most mistreated of seas. Mostly known for abuse deriving from pollution issues—with a suite of rivers from the Italian side discharging unclean waters into the northern Adriatic's paltry water volume—it is overfishing that stands out, largely unrecognised, as one of the most egregious demonstrations of human short-sightedness and greed. Fishing in the Adriatic is an excellent demonstration of human ingenuity when it comes to catching marine prey because it involves a bewildering variety of gears evolved across centuries of fishing practice. But two of them stand out for importance and impact: bottom trawls and purse-seiners, the latter mostly to catch small pelagic fishes such as anchovies and sardines. Of the two, it is the practice of trawling nets along the sea bottom that has by far the most significant responsibility for the destruction of Adriatic marine habitats. Bottom trawling is the fishing practice that brings onto Europe's dining tables the most valuable catches of a rich variety of species, including flatfishes, hakes, mullets, pandoras, cuttlefishes, octopuses, squid, and various types of shrimps, to mention just a few of them. Alas, all this cornucopia does not come without a huge environmental cost. Nets dragged over delicate bottom communities by boats endowed with powerful engines—fired with diesel fuel provided mainly by taxpayers' money and with a global carbon footprint greater than that of air travel[8]—dig their steel teeth into the sediment and destroy everything along their path. One is left to wonder whether the benefit of contemplating a roasted wild gilthead on one's plate is worth the environmental cost it entails. The Adriatic fleet of bottom trawlers numbers in the several thousand vessels, flagged from all riparian states, with Italy taking the lion's share. Each square metre of the sea floor is being swept several times per year,[9] making the Italian side of the Adriatic one of the world's most intensively trawled fishing regions. In the process, the components of the bottom fauna can't rebuild the delicate structures that are the essential constituents of the demersal ecological communities, and the sea floor is reduced to an impoverished, muddy pulp.

"Who hears the fishes when they cry?" asked Henry D. Thoreau.[10] Indeed, very few people do, or care. If we were to mow down land forests several times

---

[8] Sala E., Mayorga J., Bradley D., Cabral R.N., et al. 2021. Protecting the global ocean for biodiversity, food and climate. Nature 592:397–402. https://doi.org/10.1038/s41586-021-03371-z.

[9] Amoroso R.O., Pitcher C.R., Rijnsdorp A.D, et al. 2018. Bottom trawl fishing footprints on the world's continental shelves. Proceedings of the National Academy of Sciences https://doi.org/10.1073/pnas.1802379115.

[10] Thoreau H.D. 1849. A Week on the Concord and Merrimack Rivers. Penguin Classics.

a year to catch all the birds and mammals that live in them—the terrestrial equivalent of bottom trawling—public outcry would make such practice impossible. But underwater is out of sight and, therefore, out of mind. Add to habitat destruction the depletion of fish populations through overfishing to get a fuller idea of the damage done by bottom trawling. When this happens, the fishes are not given the chance to recover from the incurred losses. The European hake, the most economically relevant species caught by bottom trawling in the Adriatic, is caught more than five times above its replacement potential. Fishing in this way is like living on capital instead of on interest out of your bank account: little wonder if, sooner than later, we will be broke.

Bottom trawling is also one of the least selective of all fishing methods. Storage room aboard fishing boats is at a premium, and the least valuable or unmarketable catch is thrown back to sea dead or dying to make room for the prized species. During the last half-century, such discards have amounted to hundreds of millions of tonnes of wasted marine life globally. An ecocide of monstrous size which has been impoverishing the oceans and that is made possible only because of the vast subsidies the fishing industry receives from the world's governments.

If bony fishes such as hake and their similar are having a bad time sustaining the onslaught, it is the sharks and the rays, collectively called cartilaginous fishes (because they have a skeleton made of cartilage instead of bone) that are the most brutally hit. The shark's main misfortune is caused by their very low reproductive potential. In most species, females deliver only a handful of live pups—only one in the extreme case of manta rays—compared to the huge number of eggs that can be produced yearly by bony fish moms, in many cases numbering in the millions per year. A winning reproductive strategy for hundreds of millions of years, the sharks' way of reproducing has suddenly become a liability in the light of overfishing by *Homo sapiens*, and the Adriatic Sea is an excellent case in point. A comparison of cartilaginous fish landings between 1948 and 2005 revealed a decline in the overall catch rate by an astounding 94%, making the Adriatic Sea one of the world's worst places for sharks and rays to exist. Yes, you have read it correctly: 94% decline in half a century. In this case, however, it is worse than decline: eleven species—sharpnose sevengill shark, tope, angel shark, kitefin shark, sandy skate, duckbill eagle ray, common skate, rough skate, Lusitanian cownose ray, roughtail stingray, and speckled skate[11]—are no longer recorded in the region and had probably been

---

[11] The species' scientific names are, respectively: *Heptranchias perlo, Galeorhinus galeo, Squatina squatina, Dalatias licha, Leucoraja circularis, Aetomylaeus bovinus, Dipturus batis, Raja radula, Rhinoptera marginata, Dasyatis centroura,* and *Raja polystigma.*

extirpated from the Adriatic before the end of the twentieth century.[12] These extirpations mark an ominous trend of biodiversity loss in the Adriatic Sea caused by fishing.

But what about marine mammals? I had encountered bottlenose dolphins the day before, and they seemed to be doing reasonably well. In relative terms, bottlenose dolphins probably also are doing OK elsewhere in the Adriatic, but against which benchmark? What was the condition of the species, say, 100 or 200 years ago, before humans started depredating the seas with increasingly effective technologies and the unlimited availability of energy? We cannot know because, in that ancient past, the numerical knowledge to allow a robust, quantitative comparison between past and present was inexistent. Comparing our perception of the current state of affairs of any natural situation with some arbitrary point in the past, most likely within the boundaries of our own experience, is a common mistake that goes under the name of the "shifting baseline syndrome"—i.e. a situation in which over time knowledge is lost about the state of nature, because we have a hard time in perceiving changes that have occurred.[13] By doing so, we inevitably tend to lose the notion of how dramatic the deterioration has been. We would be on the safe side presuming that bottlenose dolphins in the past were much commoner in the Adriatic Sea than they are now, and this does not apply just to bottlenose dolphins but also to common dolphins, the latter once possibly even more abundant than the former. One factor that likely had dramatic consequences is that dolphins of all species in the Adriatic were the targets of systematic extermination campaigns by the fishing communities, sanctioned by the governments from all the coastal nations on the notion that these mammals are vermin that place fisheries activities in grave danger. It is hard for us today to believe that governments were encouraging fishers to harpoon as many dolphins as possible only decades ago, with monetary incentives that made such callous practices at times more economically attractive than fishing itself. Documents from the Italian side of the Adriatic describe the killing of a total of 3801 dolphins between 1927 and 1937 off the ports of Trieste, Venice, Chioggia, Ancona, Bari, and the Quarnero area (later annexed to Yugoslavia with the name of Kvarner).[14] Reasonably good records also exist from former

---

[12] Ferretti F., Osio G.C., Jenkins C.J., Rosenberg A.A., Lotze H.K. 2013. Long-term change in a meso-predator community in response to prolonged and heterogeneous human impact. Scientific Reports 3:1057. DOI: 10.1038/srep01057.

[13] Papworth S.K., Rist J., Coad L., Milner-Gulland E.J. 2009. Evidence for shifting baseline syndrome in conservation. Conservation Letters 2:93–100. doi: 10.1111/j.1755-263X.2009.00049.x.

[14] Meliadò E., Bavestrello G., Gnone G., Cattaneo-Vietti R. 2020. Historical review of dolphin bounty hunting in Italy with a focus on the period 1927–37. Journal of Cetacean Research and Management, 21:25–31.

Yugoslavia, where 335 dolphins were killed between 1933 and 1935, 278 between 1955 and 1957, and 788 between 1955 and 1960.[15] These numbers were likely just the tip of the iceberg because many animals must have escaped capture after having been struck and wounded, and died afterwards.

By the time public sentiment towards wildlife, and dolphins in particular, started to change in the late 1970s in Italy and Croatia, bottlenose dolphin populations had indeed been significantly clobbered, and common dolphins had disappeared for good. Today, the closest location where common dolphins still regularly occur is much further to the south, along the coast of Ionian Greece, where I was hoping to encounter them in the coming days.

I find such a radical change concerning human attitudes towards dolphins in the Mediterranean peoples to be mind-boggling, considering that it occurred in just a handful of decades, and that similar changes had consequences on the management policies of most of the world's coastal nations. This transformation demonstrates that the significant effort invested by sectors of civil society—chiefly the scientific and the advocacy communities—in documenting the dolphins' status and in campaigning to build up public awareness of marine conservation is far from being wasted. True, it did take at least a couple of generations to show an effect, and there is still a lot of ground to cover for marine conservation policies, for this more benevolent attitude to trickle into customary practice and become effective beyond the delimited popularity of the dolphins.

Pontoporia was advancing to the southeast with a diagonal course progressively getting her farther and farther from the Dalmatian coast, which, however, remained well visible on the port side until dark. In the last hours of daylight, I took the opportunity to make a few short naps whilst navigating in waters that were clear of obstacles, with the volume of maritime traffic just outside of the Dalmatian Archipelago still modest compared to what was expected in the waters further to the south. Around sunset, I was crossing over

---

[15] Bearzi G., Holcer D., Notarbartolo di Sciara G. 2004. The role of historical dolphin takes and habitat degradation in shaping the present status of northern Adriatic cetaceans. Aquatic Conservation: Marine and Freshwater Ecosystems 14:363–379.

a significant anomaly of the Adriatic Sea bottom, which in that portion of the region has a mean depth of about 100 m. As I could tell from the screen of my depth sounder, I had entered at that time a zone—called the Jabuka/Pomo Pit—where the sea bottom had sloped down to more than 200 m, cutting the Adriatic in half. With its tridimensional structure and slopes, the Pit contains a key habitat for several marine species of high economic importance, such as hake and scampi shrimps. This condition led the regional fisheries management organisation, the General Fisheries Commission for the Mediterranean (GFCM), to declare a Fishery Restricted Area (FRA) in the Jabuka/Pomo Pit in 2017.

As a consequence of the halt of fishing there, lo and behold, a rapid, impressive increase in fish biomass was observed, reflected by an increase in fishery catches in the surroundings beyond the boundaries of the FRA. Once again, I am made to wonder: if it is so easy to ensure that the fishers' nets catch more fish, why creating marine reserves is so difficult? The answer to the dilemma is as evident as it is depressing: to the fishing industry (unlike to the small-scale artisanal fishers), it is better to have a half-empty net today than a full net tomorrow—and if that, in the meanwhile, contributes to the destruction of the fishery, the hell with it. Industrial fishing nowadays is governed by the laws of economics instead of by the laws of ecology, and economic operations are not necessarily incompatible with the destruction of the resource they are based on if they give you the chance of eventually switching to some other more profitable undertaking.

Day 4. Adriatic Sea

Pontoporia was proceeding well under the push of her little engine across the flat seas. Sometime after midnight, the flash from the lighthouse of Vis came into sight. It was time for me to be well awake and careful, as my route was taking me to pass amongst the small islands west of Vis, with Biševo a few miles to port and Sveti Andria and Brusnik to starboard. All these small lands, somewhat secluded from human frequentation and with their steep cliffs pierced by caves, once provided suitable habitat to the highly endangered Mediterranean monk seals. Unfortunately, however, these pinnipeds likely have been extirpated from the area long ago. I hope they will re-conquer their old abodes and return here one day if they survive human pressures where they still exist, like south of here in Greece.

A couple of hours later, the flashes of the Palagruza lighthouse appeared as I was now navigating along the midline of the Adriatic, and I had to pay double attention because I was approaching the basin's main ship lane. The Palagruza islets—the largest of which was known to the ancient Greeks as *Diomedia*—were about eight miles off to the starboard side when the first light of dawn started to appear. As the sun rose from the Bosnian hills in the distance to the east, I was entering the south Adriatic, now navigating again in Italian waters. The sea here had become more similar to the rest of the Mediterranean, with waters exceeding the depth of 1000 m and with more pelagic, open-sea species like striped dolphins who, however, were nowhere to be seen on that morning. The Gargano Peninsula—the spur of the Italian boot—was by then no more than thirty miles to starboard, and the Italian coast was getting closer and closer as I proceeded, with the Apulian tableland slowly coming into view as the rising sun was painting a warm light over its outline.

It was time to converse with another companion of my maritime travels, the faithful Moka coffeemaker. I presume one has to go back to Columbus' days to find an Italian sailor that takes to the sea without a Moka, and that is only because Mokas had not been invented yet at the end of the fifteenth century, nor coffee was known to the Western peoples in those times. The Moka is a simple device made in cast aluminium with a typically octagonal section, coming in many sizes depending on how many cups of coffee one wishes to brew. However, the device's apparent constructional simplicity belies the mysterious way it works because the Moka is a temperamental contraption. It can deliver anything between the most delicious cup of coffee and an undrinkable dark slush, depending on its mood—or, to be fair, how you have been treating it and loading it with the coffee grounds. Its working principle is simple: (a) you put the water (not too much, not too little) in the bottom chamber; (b) you place the coffee grounds in the filter above (the granularity is key, and woe betide you if you compress them into the filter more than just slightly); (c) you place your Moka on the burner (the flame not too high, nor too low) and wait till the water is pushed up through the grounds as it boils, percolating into the upper chamber to treat you with the prized cup of coffee. Sounds difficult? Not really, but establishing a personal relationship with your Moka is a must if you intend to distil happiness from it. Its behaviour often seems to evade the laws of physics it is supposed to obey, degenerating into a most capricious relationship with its handler. Having established an excellent relationship with my Moka over the years, on that cool dawn, I was soon rewarded by the soft, friendly sputter announcing that the brew was ready, accompanied by that delicious fragrance of coffee able to transform most mornings into something rather pleasant.

The sight of the Apulian coast brought back the distant memory of a troubling event I had been involved in, back in 1987. I had just returned to Italy after almost a decade spent in California to complete my education as a marine ecologist. Something very wrong had occurred in the Apulian seas earlier in that winter, when 71 dolphins (of which 31 were striped dolphins, two were bottlenose dolphins, two Risso's dolphins, and the remainder unidentified because the status of the carcasses prevented species identification), 205 loggerhead turtles, plus uncounted birds and large fishes, had been found dead or dying on beaches along a modest stretch of the Apulian coastline over three months. The event had caught the attention of national media as well as the concerned authorities, and Otranto's chief magistrate Ennio Cillo appointed me and colleague and friend Bruno Cozzi, a veterinarian, as the court's experts. Bruno and I had travelled south from Milano to Otranto to perform necropsies on nine striped dolphins that the judge had confiscated in search of clues about the possible cause of such a tragedy. I still remember the nasty smell that permeated the stuffy air of the dissection room and sticking to our clothes for days (it was then that I decided that I would specialise in live, healthy, possibly happy animals freely frolicking in the seas). None of the dolphin bodies had traces of external wounds or signs of being caught in a fishing net. Nor was there any indication from subsequent analyses that any disease might have infected the animals. Instead, we found that all the dolphins had extended ulcerations in their stomachs. We interpreted this as the smoking gun of intoxication (likely by phenolic compounds, as subsequent laboratory analyses revealed) that had led to cardiocirculatory arrest and death.

In the sights of Judge Cillo was a factory in the port city of Manfredonia of the state-owned chemical giant Enichem, which had been previously caught red-handed for having committed offences involving environmental land contamination. In those years, the protection of the marine environment was still an extravagant concept, to say the least, and the Enichem factory had been allowed to legally discharge at sea enormous quantities—at the rate of 3000 tonnes every 5 days—of by-products of the synthesis of caprolactam, a precursor of nylon. Mr. Cillo had been able to verify that the composition of the sludge to be discharged at sea did not coincide with what had been declared by the company and that it contained dangerous amounts of toxic compounds, including phenols. Furthermore, there was the suspicion that the tanker carrying the chemical sludge at sea, which the crew was instructed to discharge hundreds of miles away in international waters off Libya, might, in fact, have done so routinely just beyond the horizon, a handful of miles off Apulia, to save on fuel and time. Not that discharging toxic waste at sea in one place or another would make a significant difference, but this was another sign of operational carelessness anyway. In 1988, Judge Cillo confiscated the sludge tanks and prohibited further discharge at sea. Interestingly, initial public demonstrations against the court's decision, perceived as a threat to employment and the local economy, were followed in the subsequent years by similar demonstrations; this time, however, these were aimed at the heavy legacy of toxic contamination that was left by the Enichem plant in Manfredonia, which was finally shut down in 1994.

Meanwhile, having been parachuted into the melee as incorruptible outsiders from the far North, Bruno and I had become the targets of a smear campaign by the hired guns of the "local stakeholders", who purported that we were just newbie novices who did not know what they were talking about; and that there was nothing exceptional in the massive mortality of all those marine animals, purportedly caused by a spell of cold weather. The Apulian regional government of the time seemed more concerned about assuring the salary of the 500 Enichem workers than about protecting the environment under its remit. At the same time, the industry had unleashed its minions from the local scientific intelligentsia—luckily without significant effect.

The reckless environmental policies in Italy and the Mediterranean in those times have significantly—albeit still insufficiently—improved in present days, also thanks to the conscientious engagement of intrepid members of the judiciary such as Ennio Cillo. Recalling that despicable episode with my friend Bruno—who just retired as head of the Department of Comparative Biomedicine at the University of Padua—we have at least been able to consider, with some amusement, that the smear campaign we had been targeted at the time by local villains had not negatively affected our respective careers in the least.

I was now close enough to the Italian coast to benefit from a favourable breeze that had developed in the early afternoon, blowing gently from the seaside onto the Apulian shore. Raising the sails was a welcome change, both for *Pontoporia*'s indefatigable engine and for my ears, and the boat was now progressing, silently and smoothly, towards her destination. The wind was relatively steady in direction and intensity, and I allowed myself the luxury of a few short dozing bouts under the vigilance of both radar and AIS as the coast of Italy was streaming by on the starboard side. At sunset, the breeze subsided, so it was again the turn of the engine to push the boat forward. Here, more attention was needed because the maritime traffic had increased substantively compared to the night before. When the darkness descended upon me, I found myself proceeding under a clear sky studded with stars, with the thinnest crescent of a waxing moon barely visible.

Day 5. Adriatic Sea

Pontoporia had continued in her steady progress under my supervision. However, I was looking forward to the moment the boat would have been berthed safely in the Otranto harbour, and I could have my well-deserved spell of deep, uninterrupted sleep. At dawn, the coast of Apulia was now getting closer to starboard, and I thought I could almost make out the ancient tower, about 8 miles away, which gives the name to the marine protected area (MPA) of Torre Guaceto. Looking at it from afar, nobody could have guessed that this small plot of marine territory, its surface a mere 20 km^2, is one of the most celebrated examples of marine conservation in the Mediterranean, demonstrating that the careful, well-managed protection of the marine environment is not necessarily obtained to the detriment of the local economies. Managers of the Torre Guaceto MPA, in close collaboration with both the local scientific and fishing communities, had carefully assessed the amount of fishing pressure that the area could tolerate sustainably. Then, fishing permits were assigned to the local cooperative within those limits. Results did not take long to appear, with catches stabilised in a handful of years at a level more than twice higher than that measured outside of the reserve. The caught fishes were also significantly bigger. Again, a demonstration that the wise co-management between seemingly opposing interests—environment protection and fishery—is possible, which defies my understanding of why there are not thousands of other Torre Guacetos scattered around the Mediterranean.

At long last, having negotiated with particular attention the waters in front of the opening of the busy industrial port of Brindisi, *Pontoporia* covered the remaining thirty miles separating me from Otranto under the welcome push of a moderate wind from the northwest, which had started to pick up in the meantime. Finally, I could round Otranto's breakwater and dock at one of the piers along the harbour's east side.

It was now time to do what sailors do when they hit land: sleep, relax, tour the town, meet the people, and sample the local food and wines—particularly interesting here in Apulia. In Otranto, however, there was something else which I had heard of before and wished to take this opportunity to visit. The local cathedral, named in honour of Santa Maria Annunziata, harbours one of the Middle Age's finest mosaics, created between 1163 and 1165 by a monk named Pantaleone.

So, up I went to the small hill where Otranto's cathedral is perched, not so much to ask forgiveness for my sins but to admire the mosaic. Quite frankly, I was not ready for what I was about to see, and on entering the cathedral, I was astonished by the beauty of the sight. Figures portrayed with coloured tesserae lined the entire floor of the big church. Around a Tree of Life running along the midline of the nave were representations of the Garden of Eden, of humans' virtues and vices, of the cycle of the year's twelve months, of the Great Deluge, of the Tower of Babel, and several other vignettes of antiquity's lore.

However, what enthralled me most was a fantastic admixture of animals populating the mosaic, real and imaginary, terrestrial and marine. Wherever he could find some space left on the floor, good old Father Pantaleone had managed to cram in some animal of his fancy. The resulting menagerie is a good representation of the zoological knowledge of the time, with some species well known, some less known, and some totally imaginary, but portrayed nonetheless. This observation reminded me of the early days of my marine explorations, when I knew what some animals looked like, like the dolphins, for instance, but had no idea of what other forms of life I might have encountered. In those days, little photographic documentation was available for many marine species such as beaked whales, Risso's dolphins, devil rays, and various rare sharks. But the sense of natural opulence provided by the Otranto mosaic is one that people must have felt, like Father Pantaleone, when exploring—or even imagining to explore—uncharted land or sea one millennium ago. I can imagine the excitement those ancient folks might have experienced in front of the unknown because it is the same excitement I have felt every time I have taken to the sea in search of animals. This search was often rewarded by some new discovery, which has helped push science forward by an infinitesimal increment. Adding bit after bit of knowledge, today, we clearly understand what swims in the oceans, and we can draw an uncompromising line between the existing and the imaginary. This condition has not diminished our sense of awe for the natural world, but it adds considerable reason for concern. The emerging science of historical marine ecology has reconstructed past changes in the natural conditions of the marine environment, including the trends in the abundance of fishes, reptiles, birds, and marine mammals across the past centuries, revealing the scary dimensions of the loss of nature our oceans have suffered. As far as groups of marine species are concerned, such as sharks, sea turtles, deep-sea fishes, and reef fishes, evidence was provided that levels of depletion today are around 90% of their original abundance.[16] Unfortunately, a proper, clear awareness of so profound a natural loss extended over centuries is impossible due to the mismatch between the length of the processes and the duration of human lives. And so it is that we are walking, blissfully unaware, into a future where our world is progressively impoverished.

---

[16] Lotze H.K., Worm B. 2009. Historical baselines for large marine animals. Trends in Ecology and Evolution 24(5):254–262.

4

Ionian Sea: From Otranto to Corinth

Illustration by Massimo Demma

Day 6: Otranto, Italy

At dawn the silence in the harbour was only broken by the lapping of wavelets washing against the congregation of small fishing boats moored around my vessel, as in a soft conversation, set into motion by the slight swell which managed to sneak in from the open sea. The cackle of a gull getting ready for a new day of raids reminded me that the day had excitement in store for me as well.

The time had come to bid goodbye to the Adriatic Sea, by now a familiar playground, almost a cocoon, an oversized offshoot of the Venice Lagoon, with the chromatic mildness of its pastel skies and seas and its low coastlines, thick with humanity, as far as the eye can see. The time was now to forge ahead into the vast Ionian Sea, where the bottom plunges to depths unknown anywhere else in the Mediterranean,[1] and where the waters—as Homer a bit cryptically wrote—become wine-dark. I left with a sentiment of gratitude for the good old Adriatic for walking me like a mentor from my lagoonal closet towards more oceanic adventures. Still, it now smelled too much like my backyard, and I was ready to shake myself away from it.

Having stockpiled *Pontoporia*'s storeroom with a couple of succulent pear-shaped caciocavallo podolico cheese, and with a provision of pitille, frise, pasticciotti, and similar Apulian delights from a bakery in Otranto's backstreets I had located with the help of my nose, I quickly cast off, abandoned the protection of Otranto's breakwater and eagerly raised the mainsail to the south-east.

The Ionian Sea holds significant natural treasure in the generally ill-treated condition of the Mediterranean. Bottlenose dolphins are still encountered, mostly confined to the shallower depths near the coast, but are no longer the sole non-human landlords as in the northern portion of the Adriatic. Beyond the continental shelf begins the territory of the giants, where the colossal fin whales compete for size with sperm whales; and the bizarre, deep-loving Cuvier's beaked whales, the round-headed, hoary Risso's dolphins, and, of course, the elegant, acrobatic, omnipresent striped dolphins also share their territories with them.

Unfortunately, the waters had become rather choppy on that day, whipped up by a brisk north-westerly breeze that was giving me a welcome push but was uncongenial to good sighting conditions; I knew that I had to be content with the thrill of what I might encounter, even without encounters. The idea

---

[1] The "Calypso Deep", 34 miles southwest of Pylos in the Peloponnese, reaches a maximum depth of 5,267 m, the deepest in the whole Mediterranean Sea. It could engulf the whole of the Mont Blanc, Europe's highest peak.

of the presence of wild things is often as exciting as the wild things themselves, particularly at sea where the opaque watery medium prevents one from seeing much of what happens below the surface, and anything, in the best of circumstances, beyond a depth of a dozen of metres or so.

As I was putting miles between *Pontoporia* and the Italian coast, I knew without seeing them that hundreds of metres below her keel millions of polyps of *Lophelia pertusa*—a deep water coral, white like a bone—assembled into the most fragile of coral reefs in the pitch dark of the deep slope, busily filter-feeding minute planktonic organisms and organic matter raining down on them from the upper layers of the sea. The General Fisheries Commission for the Mediterranean had declared here a protected zone—another FRA— expressly to avoid the razing of these coral constructions, which took thousands of years for the polyps to build, at risk of being destroyed in a matter of seconds by the mindless passing of a trawler's net. As I glided over this unseen and unsung natural marvel half a mile down, I could not avoid imagining the corals unwittingly awaiting their sad fate because, sooner or later, a rogue trawler would be over them, destroying their ecosystem to oblivion. Or maybe it has happened already: who is there to watch and enforce the rules? It is so difficult to ensure compliance with the rule of law in the open sea.

I had juicy plans for my visit to the Ionian. First on the list was the Gulf of Ambracia, a deep indentation in Greece's west coast inhabited by an isolated, critically endangered population of bottlenose dolphins. Then, I would travel to the sheltered waters surrounding the Inner Ionian Sea Archipelago, home not only to bottlenose dolphins but also to rare and endangered common dolphins, and where I had promising hopes of encountering the Mediterranean monk seals, also endangered. Finally, on my way to the Aegean Sea, I would sail across the Gulf of Corinth, hoping to run into its special assemblage of striped dolphins secluded there. The Mediterranean Sea is a mosaic of conditions where populations have occupied confined spaces for a long time, bringing about and conserving their genetic differentiation from their original ancestors.

Soon after leaving Otranto, the low coastline of the Apulian tableland was fading quickly beyond the horizon as *Pontoporia* proceeded briskly under the push of the Maestrale, filling her genoa sail with commanding gusts. I was crossing the Strait of Otranto, a channel separating the heel of Italy from the Balkan Peninsula, a mere 45 miles wide: the gateway to the Ionian Sea that received its name from the town I had just left.

Ah, the unblemished beauty of the open sea! With the last sight of land disappearing in my wake, I was seized by the rapture of finding myself alone with nature and allowed a much-needed dialogue with my outside world without intermediaries, facilitated by the most extreme simplification of landscapes one can find on the planet. It is not easy to describe the euphoria of forging ahead, silently and smoothly under the push of the wind, across the plane separating above from below, with no dirt in between. Alone and away from land with all its anthropic baggage of noise, smells, and nonsense. Alone but not lonely because of the perceived, albeit unseen, presence of all the marine beings hidden in the folds of the waves, possibly watching my moves. Always with a watchful eye on the sky's colour to decipher the sea's mood. And yet, at the same time, quietly ruffled in the deepest corner of my heart by a primaeval sense of unease for penetrating an open space which is perhaps a bit too open, so that the commanding sense of exhilaration is tempered by the opposite desire of returning to land—and the eye runs to the horizon searching for that distant coastline revealing that safe harbour is within reach. This inner conflict is unavoidable because, in the end, I cannot escape my nature of a land mammal, instinctively striving to seek refuge in the hole where I was born and where I can hide and rest without worrying about uncontrollable externalities.

Considering the modest pace of a small sailing vessel such as *Pontoporia*, I knew I could not reach my first destination in the Gulf of Ambracia in one day of sailing. Therefore, my plans were to anchor for the night in some sheltered bay along the way. As I approached the coast of the Balkan Peninsula, the rugged line of the mountains of Albania was emerging over the horizon on the port bow, breaking the spell of being really in the open sea. Flying fish scared by the peremptory inception of the bow were lifting off from the waves and scattering in all directions, terrified in their attempt of escaping their imaginary predator. It was a merry promenade, conducive to a happy disposition. During the hours, the Albanian skyline was streaming away rapidly on the port side, soon replaced by the hilly silhouette of northern Corfu. I had now entered Greek waters, and a shiver of excitement ran through my spine.

What is it about the sense of thrill that inevitably seizes me when I come to Greece? This feeling is certainly not unique to me. Love for Greece and her culture even has a name—Philhellenism—and has ancient roots, dating from Roman times when emperors such as Nero, Hadrian, and Marcus Aurelius, surrounded by the wider entourage of Rome's literati, openly admired the elegance of Greek intellect and tastes which so much contributed to the elevation of Roman culture in imperial times. More recently, a wave of philhellenism swept through nineteenth-century Europe in a romantic surge to support the Hellenic Independence Movement against Ottoman domination. In my case, however, I know that the source of this sense of fascination for Greece is neither literary nor romantic. It is something running deep inside, perhaps under the biochemical impetus of my genes. It is always such a joy when I happen to be travelling to Greece, an excitement that has more to do with a gut feeling of natural affiliation and an uncanny sense of having finally returned home. It is a mysterious and unexplained attraction, a sentiment dense with unknown ingredients but certainly strengthened by admiration for the physical beauty of so many Greek landscapes and the inevitable sense of awe deriving from the land where the culture I belong to was born.

Along the way, between one wave and the next, I had made up my mind that I wished to spend my first night anchored in a bay of a small island called Antípaxos. Together with her sister Paxos, slightly larger and more populated, Antípaxos is the quintessence of the small Greek island, and it seemed quite appropriate for celebrating my first night of return to this cherished land.

The sun was prepared to set as I was already well inside the "Ionian Archipelago IMMA", identified as important for common dolphins and monk seals,[2] and I was coming into the shelter created against the dying Maestrale by the large island of Corfu. As I sailed southward along the channel between Paxos and the Greek mainland, the setting sun was aspersing the continental coast with a golden light suffused with shadows, enriching the landscape with the enhancement of its tridimensionality. With Paxos now rapidly sliding behind in the starboard wake, my destination came into view. It had been a long day rocking and rolling with the wind whistling in my ears, and I was looking forward to a motionless, silent night.

Many beautiful coves are indenting the east coast of Antípaxos: I selected the southernmost one. Bright green lentisk shrubs lined the white rocks, suffusing the dusk air with aromas expelled by the scrub scorched by a long, cloudless summer day. Cicadas were slowing down the pace of their chirping as the air started to cool. No houses, boats, people, or human signs in sight. The inebriation of being alone in the quiet of the bay was palpable, the silence broken only by the repeated utterance of a blackcap—teck …teck …teck— the small bird appearing to be alarmed by something irrelevant; me perhaps.

---

[2] Ionian Archipelago IMMA: https://tinyurl.com/43m7kksb.

In the clear, darkening sky, a first-quarter moon had become visible, with the bright speck of Venus shining near it.

The sound of the anchor chain scrolling into the water from the bow had hardly dissipated in my ears when I passed out in my bunk.

Day 7: Antípaxos, Greece

The bottlenose dolphins of the Gulf of Ambracia were waiting for me, or so I liked to think, and I was anxious to oblige. The anchor was already drying in its hawse under *Pontoporia*'s bow as I was heading south-east under the push of a gentle breeze—the Maestrale having helpfully subsided—to navigate the 27 miles separating Antípaxos from the entrance to the gulf.

Seascape and landscape around conjured to transmit to the passenger an idyllic sense of peace and beauty—small villages scattered across the green Epirian hills, the farther layers of mountains culminating with the imposing mass of the Pindus in the distance, the dark blue sea—sweeping under the rug a long history of conflict and tragedy that have reddened these waters with human blood, time and again across the millennia. This region was one of the main theatres where the struggle amongst Mediterranean civilisations was consumed, since that fateful 2nd September 31 BCE when the fleet of Octavian defeated Queen Cleopatra of Egypt and her lover, the Roman consul and triumvir, Octavian's former comrade Marcus Antonius—thus setting the stage for the transition of Rome from republic to empire. The battle of Actium, right there at the mouth of the Gulf of Ambracia, was only the first of a series of confrontations fought in this region across the intervening centuries, under different banners, leaders, and portended ideals and beliefs, but ultimately pitting West against East in the struggle for Mediterranean dominance. Today, hordes of peaceful tourists from the world over, blissfully oblivious of past atrocities, flock to this area attracted by its serene beauty, to swim in its waters and sunbathe on its sands where thousands of their ancestors fought, drowned or were massacred in the name of ideals that make little sense today. Might this remain forever true?

In the approach to the Gulf of Ambracia, I was in for a surprise. I knew what awaited me because I had seen the charts, but the sudden appearance of the gulf caught me off-guard nevertheless. Nothing in the shape of the outer coastline preannounced the presence of such a large inner body of water, suddenly thrown to my face as soon as I sailed around the tip of the Preveza peninsula.

The gulf got its name from the ancient city of Ambracia, the site of which is now occupied by the modern Arta. It is also known as the Gulf of Actium, the Gulf of Arta, or Amvrakikòs Kòlpos, as it is called in Greek. It is not really a gulf; calling it an inland sea would be more appropriate. About 40 km long and 15 wide, and very shallow (with a maximum depth of 60 m), the gulf is connected to the Ionian Sea through a ridiculously narrow canal, only about 600 m wide at its narrowest point, further reduced to just over 400 m by a marina which was foolishly built exactly inside the cramped passage, further impairing the water exchange between gulf and open sea. This was a serious mistake with major consequences for the quality of the gulf's waters. At the root of this problem is another man-made blunder: the amounts of fertilisers leaching from the surrounding agricultural lands into two small rivers, the Louros and the Arachthos, feeding them into the gulf. Nitrates and phosphates composing the fertilisers cause the excessive proliferation of minute planktonic algae, making the gulf's water green and opaque. When these phytoplanktonic organisms reach the end of their life-cycle and die, large amounts of biomass sink to the bottom and are decomposed by aerobic bacteria that use up all the oxygen naturally dissolved in seawater. This would not be a particularly serious problem in a body or water undergoing normal mixing processes, such as in the open sea. Still, it is a problem in the Gulf of Ambracia during the warm season, where hot air causes the water mass to acquire a layered structure. The cooler, oxygen-depleted deep layer is heavier than the warm, oxygen-rich surface that receives oxygen by diffusion from the atmosphere. The two layers do not mix because their density is so different, preventing much of the gulf's water mass from supporting aerobic life, such as fishes. Although this condition is mitigated during stormy days, because wind disrupts the layering and mixes the waters, storms are rare in the Gulf of Ambracia during summer. By obstructing the inflow of colder, oxygenated water from the Ionian Sea, the narrowing of the connecting channel caused by the construction of the Preveza marina significantly impairs a source of oxygen, which would somewhat relieve the negative effects of a dead zone in the lower portion of the body of water.

Despite these man-made ills, the Gulf of Ambracia is a world of its own, a marine microcosm with its unique features. The gulf is also identified as an Important Marine Mammal Area—the "Gulf of Ambracia IMMA"—because of the unique population of bottlenose dolphins that it shelters.[3] The gulf's north shore is a wetland of global importance listed by the Ramsar Convention

---

[3] Gulf of Ambracia IMMA: https://tinyurl.com/2w5ywdvf.

on the protection of wetlands,[4] and occupied by extended marshes providing excellent habitat to a variety of bird species such as herons, storks, kingfishers, Dalmatian pelicans, and a rich assortment of waterfowl. Conversely, the hilly southern shore is cultivated and disseminated with the towns and villages of the Aetolia-Acarnania region.

Vonitsa, my destination, is one of these towns. A coastal village, home to a mostly agricultural community, Vonitsa has a small harbour providing decent shelter from most winds. The town also caters to a growing tourist demand, mostly of Greek provenance. Like everywhere in this part of the world, no matter how small or hidden away a place is, Vonitsa has its respectable pedigree: built on top of the ruins of the ancient city of Anaktorion founded by the Corinthians in 630 BCE, the town is dignified by a 1000 years old Venetian fortress, perching over a hill and towering above the surrounding town.

Joan Gonzalvo, who was waiting for me on the Vonitsa's pier to greet me and to help with the mooring as I was docking *Pontoporia*, is a Catalan marine ecologist who dedicates his life to the conservation of marine mammals in this part of the world, when not distracted by consuming the gastronomic attractions of his hometown, Barcelona. Vonitsa is where Giovanni Bearzi had established Tethys' headquarters in Greece for research and conservation activities of the area's marine mammals. Having entrusted in 1999 Lošinj's "Adriatic Dolphin Project" in the capable hands of Draško and his Blue World Institute colleagues, Tethys had started here in Greece a new project, the "Ionian Dolphin Project", which was initially based further to the south, in the Inner Ionian Sea Archipelago—my destination on the following day. Years into Tethys' research activities in the Archipelago, the information was gained that the Gulf of Ambracia was the best possible place to be if one were interested in investigating bottlenose dolphin ecology. So it was that shortly thereafter, after a few initial investigative forays to ascertain that, indeed, the area held extraordinary ecological interest, Tethys established a permanent research basis in Vonitsa.

---

[4] Ramsar Convention: https://www.ramsar.org/.

Joan knows the region's bottlenose dolphins, common dolphins, and monk seals better than anyone I know, and I was determined to take advantage of his expertise during my visit. Sitting in one of Vonitsa's shore tavernas, at a small table precariously placed over the beach pebbles waiting to be served a horiatiki salad, I asked Joan to give me a short update on his research there.

"We discovered that the Gulf hosts the highest density of bottlenose dolphins in the whole of the Mediterranean Sea," Joan said, "with a community of about 150 individuals permanently occupying an area no wider than 400 square kilometres".

"Bottlenose dolphins are the only marine mammal in the Gulf", he said, "probably because they are the only ones able to combine their needs with the area's specific characteristics".

"What are these specific characteristics, and how have the dolphins adapted to them?" I asked as the waitress placed in front of us a bowl with a mouth-watering mix of tomatoes, cucumbers, olives, and a thick slab of feta cheese on top, all sprinkled with olive oil and dried oregano.

"We have observed that bottlenose dolphins here have adapted well to the degradation of the water quality in the Gulf, impaired by poor circulation, fertiliser runoff from the intensively cultivated surroundings, fish farms, and urban and industrial expansion. They have modified the foraging behaviour we normally observe in this species elsewhere, typically feeding on prey living near the sea bottom", Joan said, "gorging instead on the abundant round sardinellas[5] that gather near the surface". Of course, I thought, given that the sea bottom is devoid of life due to the anoxic conditions of its waters.

During more than two decades of research efforts, Joan and his colleagues have assembled a photographic catalogue of the dolphins living in the Gulf of Ambracia, which has allowed them to derive an estimated population size of 150 individuals.[6] DNA analyses performed on bits of the dolphins' skin remotely collected with minuscule darts have confirmed that Ambracian bottlenose dolphins belong to a distinct population, with very limited gene exchange with their relatives in the open Ionian Sea—although a few amongst the most adventurous individuals from the Ambracian population had been observed happily swimming all the way south to the Gulf of Corinth, 265 km away, and some even up into the Northern Adriatic almost 1000 km away, off the coasts of Slovenia. The Ambracian dolphins' reduced population size

---

[5] Round sardinella, *Sardinella aurita*.

[6] Gonzalvo J., Lauriano G., Hammond P.S., Viaud-Martinez K.A., Fossi M.C., Natoli A., Marsili L. 2016. The Gulf of Ambracia's common bottlenose dolphins, *Tursiops truncatus*: a highly dense and yet threatened population. In: G. Notarbartolo di Sciara, M. Podestà, B.E. Curry (Editors), Mediterranean marine mammal ecology and conservation. Advances in Marine Biology 75:259–296. https://doi.org/10.1016/bs.amb.2016.07.002.

combined with their isolation caused the population to be assessed as "critically endangered" in the IUCN Red List of Threatened Species,[7] thus emphasising its extremely high extinction risk.

The main reason why marine ecologists view the bottlenose dolphins of Ambracia like kids in a candy store is that the dolphins here live in a simplified ecosystem that is more amenable to scientific investigation than most other places. First, the area is contained—think of it as a very large swimming pool—where the animals live in isolation from the outside world (the few individuals who were found to be wandering outside are irrelevant to the big picture). Second, the dolphins are sufficiently few to allow the researchers to know most of them, one by one, including their sex and, to some extent, degrees of inter-relatedness. Third, human–dolphin conflict in the gulf is minor, considering that the round sardinellas the dolphins feed upon are of minor economic interest, and conflicts are significantly reduced compared to most other places where dolphins and humans must coexist. Finally, the food web the dolphins sit on top of is a simple one, with a few steps separating its components: planktonic microalgae, zooplankton feeding on the algae, small fish feeding on zooplankton, dolphins feeding on small fish. Whilst admitting that the above description is an oversimplification, because things in the real world always turn out to be more complex than they appear at first glance, the gulf nevertheless provides a textbook case to describe the relationships between a marine top predator and its environment. A science treasure throve that is still, to a large extent, untapped by Joan and colleagues, who have been until now predominantly concerned with investigating conservation issues.

"So what do you think will be the future of Ambracian bottlenose dolphins?" I asked, "How secure is their survival here, given all these factors at play?"

"Despite the appearance of a dolphin population living at high densities in an ecosystem where there seems to be plenty of food and relatively little human conflict and disturbance", Joan said, "the issue of environmental quality of the gulf and its waters needs to be seriously addressed".

"What are the most urgent measures you suggest should be taken?" I said.

"The matter mainly concerns ecosystem functioning, but there are also health concerns deriving from the poor quality of the gulf's waters and its impaired circulation", Joan said. "For instance, we have noticed more than once the onset of various types of skin diseases on the dolphins' bodies".[8]

[7] Gonzalvo J., Notarbartolo di Sciara G. 2021. *Tursiops truncatus* (Gulf of Ambracia subpopulation). The IUCN Red List of Threatened Species 2021: e.T181208820A181210985. https://doi.org/10.2305/IUCN.UK.2021-3.RLTS.T181208820A181210985.en.

[8] Gonzalvo J., Giovos I., Mazzariol S. 2015. Prevalence of epidermal conditions in common bottlenose dolphins (*Tursiops truncatus*) in the Gulf of Ambracia, western Greece. Journal of Experimental Marine

"Actions would ideally be framed within the future management plan for the area", continued Joan. "The protection regime created by a Natura 2000 designation provided by European law,[9] which had been previously earmarked mostly to protect birds in the marshlands along the northern shore, was recently extended to the whole of the gulf to support the conservation both of bottlenose dolphins and loggerhead turtles, also common here".

"So, do you think that the legal framework in place could be sufficient to address the dolphins' conservation problems?" I said.

"Yes, but only in theory", said Joan. "It is compliance with rules which is at fault. Greece is slowly recovering from a hard economic crisis, and Aetolia-Acarnania is one of the country's most depressed regions. Agriculture and fish farming come before the dolphins' welfare in people's minds. Goodwill cannot always be made to coincide with reality - and observance of the rule of law never was a typical Mediterranean virtue anyway".

Having docked *Pontoporia* in Vonitsa's small harbour, I jumped aboard Joan's nimble inflatable to make a quick foray into the Gulf of Ambracia to meet the dolphins themselves. Gliding at high speed, in minutes we were in the middle of a gulf that looked more like a lake than a sea, with its glassy waters surrounded by a coastline hardly discernible in the midday haze. It did not take much to find what we were looking for. True to expectations, a group of six dolphins appeared in the distance, splashing around at the surface and creating havoc amongst a school of round sardinellas, with flocks of gulls and terns noisily joining in the excitement. It was a merry sight if you could avoid watching it from the sardinellas' point of view. Trapped against the surface from below, the fishes got short shrift from the living torpedoes zooming underneath. The excited birds were taking advantage of the sardinellas desperate enough to jump out of the water. It was an extermination operation conducted with military precision.

The dolphins ostensibly were not minding us being there close to their doings, and watching them. Not unlike their close relatives in the Kvarnerić, I had the impression that they knew well Joan's boat from the noise it made

---

Biology and Ecology 463:32-38. https://doi.org/10.1016/j.jembe.2014.11.004.
[9] For more information about this Natura 2000 site: https://tinyurl.com/36sv4ewy.

with the hull and the propeller, and tolerated his expert and respectful manoeuvring by going about their business as if we didn't exist. It was their home; we were the guests, and they were in control.

It is always an emotion to be with the dolphins, no matter how often I have seen them. But there is something special about bottlenose dolphins that makes them particularly congenial. Somewhat portly and with a lacklustre greyish attire, their looks are not particularly striking compared, for instance, to common dolphins with their slender silhouette, or striped dolphins with their elegant flank design. But when bottlenose dolphins swim around you, they convey an endearing sense of determination, intelligence, and self-confidence that is unique to them. Often wary of boats, if they keep at a distance it is because they know what there is to be known about humans. By contrast, when they approach you it is because they have decided they can trust you, in a flattering show of acceptance.

People have been exploiting human fascination for dolphins to make money. Still to this day, dolphins—in fact, mostly bottlenose dolphins—are being captured from the wild in Japan, the Solomon Islands, Cuba, and a few other places, and confined for life in concrete tanks so that the wide public can be made to watch them at close quarters and interact with them for a price. Unfortunately, it is impossible to replicate the living conditions essential to the well-being of dolphins in these contemporary equivalents of the ancient menagerie concept, where wild animals are exposed to humans who would not have opportunities to see them otherwise. The animals need space, a choice of companions, and, ultimately, freedom that they cannot get in a concrete tank. There is no good reason for allowing such practices today except for fattening the wallets of unscrupulous practitioners.

Keeping dolphins in captivity is still good business in many parts of the world, and business is hard to kill with policies. So it is that an incipient marketing decline, caused by a growing sentiment of discomfort amongst the public about watching symbols of freedom withering in a cramped tank, is counteracted by conquering the most gullible through lies. "You see, dolphins smile because they are happy in human custody" (they do not; it is the way their faces are anatomically made); "they can help to heal children with problems" (you can do it just as well with dogs or ponies—and even with these animals effects have yet to be convincingly demonstrated). "They are the ambassadors of their kind to humans" (there is nothing educational about watching a dolphin suffer deprivation of its freedom and natural environment). "Behavioural science is facilitated by captivity" (just try to provide a good description of normal human behaviour based on observing inmates serving life sentences).

As I returned to Vonitsa after the encounter with the exuberant Ambracian dolphins, high spirits from the experience were tainted by a sense of bitterness. Too many dolphins worldwide are still languishing in concrete tanks for the wrong reasons; about 2000 of them, based on a recent estimate.[10] I remember well, as a teenager in the early 1960s, the first time I saw a captive dolphin in a tiny pool in a travelling circus in Milan. I was terribly excited by the close contact with the animal, but the situation felt so wrong that it seemed to me that my excitement contained an element of sin. Admittedly, the sense of wrongness for dolphin captivity has taken a few years to sink fully but irreversibly into my conscience. Today, crowds still flock to dolphinaria to get excited about the dolphins. How long will it take for the whole of humanity to see these establishments the way they really are, that is, prisons for innocent beings serving a life sentence?

Day 8: Vonitsa, Greece

The calm weather continued on the new day. The Maestrale that had given me the vigorous, welcome push from one side of the Ionian to the other two days ago had now definitely gone to sleep, and, for my progress, I was once more in the hands of *Pontoporia*'s auxiliary engine.

Small summer cumuli were building up in the distance as the morning air got warmer, mirrored by the flat sea and festooned in perfect duplicate above and below the horizon. The calmness of the day was contagious to the spirit. I have an extreme fondness for calm weather at sea, perhaps derived from my lagoonal upbringing. Moving across a flat sea, not unlike walking across a field covered by untouched snow, gives me a sensation of physical well-being and comfort, which the body responds to with a pleasant tingling effect along the spine. But I know well that the bliss provided by a flat sea can quickly morph by the arrival of the first gusts of a storm, and the spell can soon be replaced by a sense of disquiet when I get shaken to and fro by the waves stirred up by that storm. I cannot help but feel betrayed by a sea that has converted from calm to stormy, transforming itself from being a friend into something I must fight against. Personal relations with Poseidon's capriciousness aside, a calm sea also strongly increases the chances of making contact with marine wildlife. One can see animals that break a calm surface far away, hear the cries of the birds, the jumps of the fish, the powerful blow of a whale and even the punier one of a dolphin. For a marine naturalist, a calm sea is an open book that is being closed when it gets rough: this is no small feat, and for a naturalist a huge fault of a stormy sea.

---

[10] Here is a recent estimate of the number of cetaceans held in captivity: https://tinyurl.com/ykc5z5va.

Conditions on that day were perfect for a foray within the Inner Ionian Sea Archipelago, a collection of various sizes of islands strewn along the Greek mainland's west coast, sheltering a stretch of coastal waters inhabited by all kinds of fascinating beings. The north-to-south string of the larger islands— Lefkada, Kefalonia and Zakynthos—encompasses an inner sea scattered with a quantity of smaller lands—from the most famous of them all, Ulysses' Ithaki, to Arkoudi, Atokos, Kalamos, Meganisi, Skorpios, and many other smaller, famous or obscure, inhabited or uninhabited, named or unnamed islands, islets and rocks.

In particular, by visiting the Inner Ionian Sea, I had the not-so-secret hope of encountering two of the most iconic and charismatic Mediterranean marine mammals: common dolphins and monk seals. Both once widely spread across the Mediterranean, they have become so rare in the past decades that they are no longer easy to find anywhere except in specific areas—such as here.

The trip from Vonitsa to the Inner Ionian Sea was not a long one. Once outside of the Gulf of Ambracia, I had the choice of transiting across a narrow channel between Lefkada and the Greek mainland, where a sliding bridge periodically opens to allow the passage of boats; or to give Lefkada a wide berth by navigating again across the open waters to the west, and coming back inside from the south. I selected the second option because I was not sure when the sliding bridge was being opened. Joan was to leave later with his research inflatable, travel along the faster route across the channel any time he wished because he could pass under the bridge even if closed, and meet me in the Inner Sea in an area where monk seals had often been seen lately. I found the whole plan to be terribly exciting.

Back in the open sea, my eyes clicked inadvertently into scanning mode. Sighting conditions that day were excellent, and bumping into something exciting, such as a fin whale, a Cuvier's beaked whale, or even a spinetail devil ray, was not out of the question. Unfortunately, I did not see much that day in those pelagic waters. That does not mean that there was not anything to be seen. Chance greatly influences what one encounters when navigating, particularly when scanning the horizon from the low deck of a small vessel, as was my case. The sensation of extending one's sight over great distances, even in perfect sighting conditions, is a mere illusion. To understand this, try to plot on a chart of the entire Ionian Sea a circle having a radius of a couple of miles—which is as far as you can get with your eyes, in the best of cases, from the deck of a small vessel such as *Pontoporia*. You will see your tiny circle drowning into the immense space you are navigating. So it is that encountering animals that naturally live at low densities, as top predators like cetaceans do, is, to a large extent, a matter of luck. This is why it so often happens to cross seas, even those known to be rich in wildlife, without encountering large

Fig. 4.1 Breaching in the calm waters of the Inner Ionian Sea Archipelago, Greece, this common dolphin displays the species' typical flank hourglass colouration and the long, pointed rostrum (Illustration by Massimo Demma)

marine fauna. To complete the picture, one should also consider that marine mammals tend to hide under the surface by diving for extended durations, and that some of them deliberately avoid being around humans and dive when hearing the noise of an approaching engine.

My navigation that day was uneventful until I reached the southern tip of Lefkada and turned east to enter the Inner Sea across the passage between Lefkada and the northern tips of Kefalonia and—just a bit more inside— Ithaki. The first island I found was Arkoudi, the coast of which I hugged to the south until the steep profile of a second island, Atokos, came into view.

It was during the approach to Atokos that I was cheered up by the first memorable encounter of the day. They appeared at a distance, first a tell-tale commotion at the surface and then pointed dark fin tips cutting through the water with speed and determination. A small group, maybe a dozen common dolphins[11] were swimming towards my boat, and soon, a handful had positioned themselves under *Pontoporia*'s bow, eagerly riding the pressure wave created by the boat's progress. After hours of contemplating water, luck returned to me again (Fig. 4.1).

---

[11] The common dolphins, a cosmopolitan cetacean, was described by Carl von Linné in 1758 with the name *Delphinus delphis*. The Mediterranean subpopulation is listed as Endangered: https://tinyurl.com/3brrkpwt.

Sometimes, zoological nomenclature seems to have a wicked sense of humour. A swimming oxymoron, the common dolphin—in the Mediterranean Sea, at least—is not common at all. There are many places in the world where the species is indeed abundant, which was once also the case in the Mediterranean; sadly, no more. Sometime during the second half of the twentieth century, something went wrong and common dolphins practically disappeared from most of the region, except for a few pockets of resistance, such as in the Alborán Sea in the extreme west, in the Strait of Sicily, in parts of the Aegean Sea, off Israel, and indeed here, in the Inner Ionian Sea Archipelago. Nobody has yet figured out what happened to Mediterranean common dolphins. As discussed when I was sailing south across the Adriatic Sea, their disappearance from there can be at least in part explained by the extermination campaigns carried out by the national governments in the twentieth century. However, similar mindless policies having not been adopted throughout the Mediterranean, government-sanctioned culling is an unlikely explanation for the overall decline of common dolphins in the region. We might never understand what happened to the species in the Mediterranean. Systematic studies of cetacean ecology in the region began no earlier than in the 1980s, and whatever happened before is clouded in the thickest of fogs, which prevents comparing conditions across time and understanding phenomena. Quite frustratingly, any intervention to reverse the negative trend of Mediterranean common dolphins will remain a challenge if we do not know the cause of their decline—although the case of the common dolphins in the Inner Ionian Sea provides interesting insight into the conundrum, as we shall soon see.

People who say "dolphins" when they see some at sea often display little appreciation of how different the various dolphin species are from each other. The comparison between bottlenose and common dolphins—different from each other no less than donkeys are from zebras—is a case in point. Whereas bottlenose dolphins are large, stocky, and greyish, common dolphins are smaller, slender, and adorned by a most elegant attire, with a dark back, an ochre chest patch, a grey tailstock and a white belly, their colours being distinctly separated in an hourglass pattern on the flank. Whilst bottlenose dolphins, confident of their might, like to display brash and direct behaviour, common dolphins have a more evasive demeanour, swiftly appearing out of nowhere and always seeming on the verge of disappearing again. And yet, whereas common dolphins seem unable to resist flocking in front of advancing vessels to play with the bow wave, wiser bottlenose dolphins only engage in the sport if they feel safe about getting too close to humans. The gut feeling I get when observing the two species is that bottlenose dolphins are cast in iron, adaptable, smart, survivor types, able to make the best of everything around them. By contrast, common dolphins seem blown in glass, delicate, almost fragile, and less adaptable to change. One is tempted to start from such a comparison when trying to understand the causes of the common dolphins' observed decrease in the Mediterranean.

True to predictions, the small school of common dolphins that had greeted me at my entrance in the Inner Ionian Sea did not stay around for more than a few minutes, and vanished in the deep all of a sudden, only to reappear in the distance minutes later in my wake, small dark fin tips protruding from the surface as they swam away. Nevertheless, I hugely appreciated their greeting gesture, as their beautiful silhouettes persisted in my recollection long after they had vanished.

Soon I was under the stunningly beautiful cliffs of Atokos' south coast, the white rock composed of thin layers of limestone warped by geological upheaval, and reflected in the blue waters in a breath-taking interplay of colours. Atokos' cliffs are not only beautiful: their soft carbonate texture is conducive to the presence of underwater caves, which create an ideal monk seal habitat and are still inhabited by the endangered pinniped. I was finally entering the seals' world, where the actual animals are still present in numbers, and the likelihood of catching a glimpse of one out of the corner of my eye was real.

Things were looking even more promising near the tiny island I was heading to next, where I had given my rendezvous with Joan. Formikoula is a small, uninhabited islet close to the southwest tip of the island of Kalamos, slightly more than 500 m wide from tip to tip along its greatest diameter and known to be frequented by monk seals. It was there, nestled in a pretty cove along the east coast of the islet, in waters coloured in bright turquoise, that I found Joan already anchored with his inflatable boat. As he came aboard to help prepare a light lunch based on *tyropites* and *spanakopites* fresh from his preferred Vonitsa's bakery, I had the opportunity to quiz him about the Inner Ionian Sea's marine mammals and learn more about the common dolphins' plight.

"The presence of common dolphins in the Inner Ionian Sea was the very reason why the Tethys Research Institute started a project here in Greece in the 1990s", said Joan. "In those years, it had become apparent that the Mediterranean population of common dolphins was doing badly, and the discovery of an area where the species was still abundant triggered the decision to establish a research station here".

"Why do you think common dolphins liked being around the Inner Ionian Sea Archipelago?" I asked.

"Whereas the area's bottlenose dolphins, unlike their peculiar relatives in the Gulf of Ambracia, were interested mostly in preying on demersal fishes near the coastline and are busy patrolling the surroundings of the fish farms where there is often something to snatch", said Joan, "common dolphins were doing quite well within the Archipelago because this was a breeding ground for small schooling fishes such as sardines and anchovies, which are the main components of their diet here". He then added, "it was not uncommon to see large schools of these graceful mammals darting around at the surface and gorging on the small fishes, amongst a mayhem of screaming birds eager to join the banquet".

In those years, the area was only exploited by a modest number of small-scale artisanal fishers, targeting bottom species such as breams and mullets with their trammel nets. These fishers were not having a significant impact on common dolphins. Things changed radically when the industry stepped in, barging into the area with a dozen large vessels specifically to target sardines and anchovies. These vessels were catching the fishes by casting their nets—called purse seines—around the schools at the sea surface and hauling them up in one go. With regulators conveniently turning a blind eye, it did not take long for sardines and anchovies to become severely depleted in the Inner Ionian Sea.

"By way of example", said Joan, "during the traditional summer 'sardine festival' celebrated in the nearby coastal town of Mytikas, which contemplated the frying of sardines in a gargantuan pan in the public square to the bystanders' delight, at some point, sardines were no longer available and had to be replaced with sea breams provided by nearby fish farms".

"Ironically, it had become a 'sardine festival' without sardines. You can imagine what this egregious example of mismanagement had on the common dolphin population", Joan said. "We knew how many dolphins there were in the area because, during the years of our investigations, we had collected a catalogue of 143 individuals identified through high-definition photographs of their highly distinctive dorsal fins, showing characteristic white patches and the most minuscule nicks and notches. By 2007, we could tell that in 13 years, their presence in the area had collapsed by 90%.[12] We were dismayed

---

[12] Bearzi G., Agazzi S., Gonzalvo J., Costa M., Bonizzoni S., Politi E., Piroddi C., Reeves R.R. 2008. Overfishing and the disappearance of short-beaked common dolphins from western Greece. Endangered

at seeing the dolphins disappearing from our study area and almost incredulous at having been able to witness such a disaster in our lifetime, let alone in 13 years. But we did not understand what happened to the dolphins: we couldn't believe they had starved to death and hoped they had just moved elsewhere".

In successive years, casting their net wider to more northern waters, Joan and his colleagues extended the surveyed area to the north, all the way to Corfu. And, guess what: they found "their" common dolphins again. They knew they were the same animals they had seen in the past in the Archipelago because they could recognise many of their old acquaintances by comparing photographs of their body characteristics.

"What had happened was that, because of the drastic reduction of prey resources in their preferred habitat, the dolphins were forced to increase their range to find the food they needed to survive", Joan said. "We occasionally see them here in the Archipelago, but sightings have become extremely rare. Today, they are forced to travel more to find their food".

It is still too early to tell to what extent this change has affected their survival. A greater energy expenditure to look for food over a wider area, for example, may not be compensated by new prey types, particularly if these have a lower nutritional value, ultimately causing the population to decline.

Here in the Inner Ionian Sea Archipelago, the smoking gun for the decline of common dolphins was prey depletion by overfishing. In the northern Adriatic, it likely was extermination attempts by the governments. What had happened everywhere else in the Mediterranean? Here, at least, we know what should be done to address the problem: manage fisheries sustainably, something all the world's countries have agreed to do anyway for their own sake— not necessarily for the dolphins—but unfortunately also something that too few countries have managed to accomplish effectively so far.

"Change" is the name of the game here, like everywhere else I was going. Humans are sufficiently long-lived to be able to witness a change in the natural systems they live in during their lifetimes. Still, the environmental change rate is accelerating so much that we can see it happening in much smaller time windows—like in the decade Joan and his colleagues had been monitoring common dolphins in the Inner Ionian Sea Archipelago. Change is nowhere as obvious as in the behaviour of wild animals, as they struggle to adapt to the continuous shifting of goalposts caused by humans, either directly, like in the case of the overfishing of sardines, or in more oblique ways, such as by warming up the world through the belching into the atmosphere of carbon from

Species Research 5:1–12. doi: 10.3354/esr00103.

cars, planes, homes, farming, and factories—or through the engine exhaust of bottom-trawlers, as we had seen when crossing the Adriatic Sea.

Change, however, does not always need to go from good to bad and from bad to worse. Rare instances exist, even in our wretched times, when a change in nature has a positive sign. When this happens, those who fight for a more sustainable human influence on the planet get a tremendous boost of optimism. Take, for instance, the case of the Mediterranean monk seal.

Unlike the cosmopolitan bottlenose and common dolphins, Mediterranean monk seals, as their name implies, are exclusive to the Mediterranean region (although two colonies also exist in the Northeast Atlantic Ocean: a large one in Mauritania and a smaller group around Madeira). There is an aura of mystery surrounding the species, which starts from the reason why it was named "monk seal" by Johannes Hermann when he first described it for science in 1779.[13] Some say because of the seals' hermitic behaviour of frequenting secluded caves; some because of their brown colour resembling a monk's habit; some comparing the seal's double chin to that of a corpulent abbot. I think that we should more honestly admit to not having a clue as to why Hermann named them monk seals in the first place; regrettably, in the old days, it was common for zoologists to omit explanations of why they had chosen a particular name for the species they described. Once widespread across the Mediterranean and Black Seas, in recent centuries, Mediterranean monk seals have been reduced by incessant human persecution to a few hundred individuals, presumed now to be in the region of the 600s–700s. Human encroachment of their habitat and conflict with fisheries have combined to extirpate them from the western Mediterranean and the Adriatic. However, wandering individuals are occasionally seen throughout their former range, including in Italy, Spain, and various Eastern Adriatic and North African nations.

Although habitat degradation and disturbance by humans, as well as accidental drowning in fishing nets, have concurred to undermine the seals' well-being and survival, the number one cause of the Mediterranean monk seal decline has been deliberate killing. Exasperated by competition with a marine predator clearly superior to them, small-scale artisanal fishers in the Mediterranean have kept monk seals in their sights whenever they had the chance. Whilst competing with human fishers for decreasing prey, seals inevitably end up damaging their nets and catches, putting a strain on the fisherfolk's often marginal economies: a situation begging for retaliation. The

---

[13] Mediterranean monk seals were described by Johann Hermann in 1779 with the name *Phoca monachus*. The species was later classified under its own genus *Monachus* by John Fleming in 1822.

marginality of these economies is caused by a combination of factors, such as egregious fisheries mismanagement at the national and local levels and the tolerance of illegal fishing practices leading to resource depletion. This condition has very little to do with monk seals, who, however, fit conveniently in the role of scapegoats. In the end, the conflict between small-scale fishers and monk seals has taken the shape of a battle of the have-nots where everyone is the loser, including the broader society that is being denied a healthy natural environment, rich in biodiversity and charismatic species.[14] On top of it all, and quite frustratingly, most laypersons in Mediterranean societies are blissfully ignorant about the very existence of a Mediterranean seal, sadly at the root of the absence of public concern for the species' fate. So it is that the disappearance of Mediterranean monk seals from the face of the planet, still far from being averted, deprives them even of the honour of public grief, as doubtful as such honour might be.

As a flagship marine mammal of the Mediterranean region, one would expect that the institutions would have given special consideration to its conservation and welfare. Admittedly, the species is today considered a conservation priority by European law. Unfortunately, at the national level, this could not be farther from the truth. Once abundant on all shores, from Gibraltar to Egypt, monk seals have been the victims of relentless persecution by fishers, who see these mammals as competitors and a nuisance, with governments mostly turning a blind eye to the killings. The result is that all the seals surviving at this moment in time in the Mediterranean Sea are mostly concentrated in the Aegean, the Anatolian shores of the Levantine Sea, and the western Ionian Sea. And yet, for reasons that will be further explored during my visit to the Aegean Sea in the coming days, the tide seems to be finally, albeit timidly, starting to turn in the seals' favour.

"My story with the monk seals is a love story", Joan tells me when I mention my wish to have at least a glimpse of the elusive mammal. "I felt so jealous when listening to stories from Spanish colleagues who had had the chance to see many monk seals when studying the substantial colony living along the shores of Mauritania, out on Africa's Atlantic coast. Having started to work here in Greece in 1998, it wasn't until 2012 that I had my first encounter with a monk seal. I remember well that episode: my hands were shaking, my voice was trembling, and I was covered in sweat as if I had just found in front of me the Loch Ness monster".

---

[14] Notarbartolo di Sciara G., Kotomatas S. 2016. Are Mediterranean monk seals, *Monachus monachus*, being left to save themselves from extinction? Advances in Marine Biology 75:359–386. https://doi.org/10.1016/bs.amb.2016.08.004.

At that time, the researchers from Tethys, who were concentrating on dolphins, were rather ignorant about the seals' biology; even telling a male from a female was a challenge. One of the first thoughts that impressed them when contemplating the animals casually swimming in front of their boat was how vulnerable they were to being harmed by the ill-intentioned.

"Since that day of my first encounter, we started observing more and more seals", continues Joan. "This was a life-changing experience not only from the professional point of view, as seals progressively became our study subjects in their own right, just like the dolphins. It was also an emotionally loaded experience because we witnessed a comeback that we never hoped to see in our lifetime".

Granted, Mediterranean monk seals continue to be endangered, and it is far too early for any sentiment of complacency about their recovery. Fishing still occurs in the area, at lower intensities, perhaps, and killings are still reported from time to time. Perhaps more threateningly, tourism keeps growing, and the potential for disturbance in summer, a delicate phase of the monk seals' year when they mate and give birth, is increasingly significant.

"In recent years, during the peak of the tourist season", said Joan, "the word had spread that Formikoula was the place to be if one wished to see seals or even be in the water with them, and tens of boats of all sizes and shapes were converging here every day to enjoy the unique experience. Of course, I cannot blame them, given the excitement that sighting a monk seal in its natural habitat still elicits in me so many years after I made my first encounter. However, most tourists don't seem to realise how endangered monk seals are, and some of them engage in behaviours that cause significant disturbance to the seals. Last August, we noted a newborn seal pup on the beach inside one of Formikoula's caves, but later in the day, a bunch of tourists entered the cave, and the next day – not surprisingly - the pup was no longer there. Since pups cannot swim for several days after their birth, we fear that the unwitting intrusion by the tourists in the cave might have had a lethal effect on the poor little seal".

Nowadays, monk seals are now seen with increasing regularity in the Inner Ionian Sea. Whilst in the first 14 years of their activities in the area, Tethys' researchers collected less than 20 monk seal sightings, in recent years, when surveying the Inner Ionian Sea Archipelago, the animals are sighted almost every time, occasionally up to 16 individuals huddling together inside a cave. Seeing monk seals in these waters has become easier today than seeing common dolphins.

"We no longer think that sightings of more and more seals can be due to chance", said Joan, "either the number of monk seals has increased here in recent years, or they have become more confident and are allowing themselves to be seen by us. Most likely, a combination of both".

Silence settled in as Joan terminated his story, and I contemplated with expectation the perfectly still waters of the small bay we were moored in. The bay was surrounded by low, brownish cliffs covered by lush vegetation. Crowds of marine birds were perched on the green shrubs, including a cluster of yellow-legged gulls standing in a non-belligerent arrangement with an equal assemblage of European shags. Even the birds, the gulls in particular normally known as loud characters, were dozing in the warmth of the afternoon, the silence only pierced by the drawn-out twittering of alpine swifts[15] racing at breakneck speed inches from the island's cliffs. The islet's coastline had an irregular profile strewn with indentations, each of them lined by small pebble beaches. In between, the lower portion of the cliffs at water level was riddled with caves of various sizes, and small beaches could be seen also inside the caves. Altogether, this seemed the perfect monk seal habitat.

Other boats that had made a stop in the bay earlier during the day had left. We were alone, and the quiet was absolute. Wouldn't it be nice, I thought, if a monk seal were now to swim across the bay, testifying with her presence the solidity of Joan's story? I was scanning the water intently, trying to remember from my only previous experience in Patmos, decades ago, what a swimming monk seal would look like. And indeed, a first seal soon appeared at a short distance from *Pontoporia*, materialising as by magic out of nowhere, probably emerged from one of the many underwater caves the islet's rocky bottom is scattered with, gruyere-like. The animal was swimming lowly, seemingly quite relaxed, the prominent round head turning left and right as if uncertain what way to go—so different from how dolphins proceed, who always seem to know exactly what to do. The seal was not very big and had a greyish back, so that it might have been an adult female or a young male; adult males are much larger, and their back is almost black. She came within about 10 m from *Pontoporia* so that I could see quite well the details of her head as she was

---

[15] Yellow legged gull, *Larus michaellis*; European shag, *Phalacrocorax aristotelis*; Alpine swift, *Tachymarptis melba*.

gliding effortlessly along the surface with her vigilant eyes fixed on us, unworried, before diving and disappearing from sight. A few minutes later, a second seal appeared from a different part of the bay, heading where the first seal had disappeared and diving in the same spot. We did not have to wait long for both of them to surface again, as they started to play together like two dog puppies, totally oblivious of our presence metres away.

Joan was smiling, noting my transfixed state, as he getting ready to return to Vonitsa for the night. So this area was indeed one of the last monk seal paradises left on Earth, like I was told. I prepared for a night's rest because a long navigation awaited me the following day. As I cuddled in my bunk to sleep, monk seals were all around me: in my dreams for sure, but quite likely also in reality, across the thin frame of *Pontoporia*'s hull, busily looking for some octopus-based dinner.

Day 9: Formikoula, Greece

The day ahead was going to be a long one, as I wished to reach the harbour of Corinth before nightfall. It was again a windless morning, one in which I had to avail myself of *Pontoporia*'s engine to progress. This meant up to 15 h of motoring, considering the boat's cruising speed, although I was also counting on a breeze coming up later in the day to fill my sails and push me forward faster and more comfortably.

This would be my last leg in the Ionian Sea, which had been so generous regarding natural emotions so far. As the coast of the Acarnanian region streamed along the port side, I hugged a small group of islets called Echinades, ending up in the Gulf of Patras, where I was able to point *Pontoporia*'s bow to the east, forging ahead into the Gulf of Corinth.

Again, as I was navigating along the long crack separating the Greek northern mainland from the Peloponnese, I could not avoid thinking about the wars between East and West that have ravaged this region over and again in the past centuries. I was particularly struck by the contrast between those hostile times of the past and a most peaceful present—and considered how lucky I was to light-heartedly decide to take to the sea in a small boat, alone and unarmed, engrossed in observing the manifestations of nature without the faintest worry about being attacked by pirates, barbarians, bandits, fanatics, enemies, or invaders. The contemplation, understanding, and appreciation of nature, so essential in a time in which nature is being destroyed by the economic strait-jacket humanity is stuck in, is unfortunately incompatible with war and unrest.

The low-lying coast and wetland I was admiring to my port side had been the theatre of one of the bloodiest episodes of the Greek War of Independence in the 1820s, when the joint Ottoman and Egyptian armies managed to capture and destroy the city of Missolonghi after a series of sieges, the third and final of which, in 1825, lasted 1 year and ended with the massacre of thousands of civilians, women and children included. Further to the east, less than 20 miles from Missolonghi, the small port town of Nafpaktos—known to Venetians as Lepanto—reminded of another epic confrontation in those same waters two and a half centuries earlier, when the fleets of the Holy League, a coalition of European Catholic states promoted by the Pope and led by Venice and Spain, defeated the fleet of the Sultan. The naval battle of Lepanto, in which almost 40,000 lives were lost, marked a historical turning point whereby Ottoman military expansion into the Mediterranean started to decline.

The strait's narrowest point is slightly wider than a nautical mile and separates the towns of Rion in the Peloponnese from Antirion on the northern mainland. Since 2004, the two locations have been connected by one of the world's longest cable-suspended bridges. As I navigated across the 60-metre-tall passage under the imposing construction, whilst buses and lorries thundered above me, I left the Gulf of Patras and made my entrance in the Gulf of Corinth, known in medieval times as the Gulf of Lepanto.

I did not pass under the bridge without admiration for this large, white, stunning man-made construction spanning the water and allowing crossing from one land side to the other in minutes. I remember well when travelling by land across the region in earlier years, having to laboriously negotiate the strait aboard old, rusty ferries. As I transited under the bridge, I wondered what marine animals thought when one day they woke up and had to swim under the Rion-Antirion bridge in their routine peregrinations. The question of whether dolphins and other marine animals cease to swim across straits if these are joined by bridges was the subject of debate in Italy years ago when the government had embarked on a questionable plan to build a bridge across the Strait of Messina, to join Sicily with the Italian mainland in Calabria. Opponents of the bridge's construction based on environmental concerns had argued that the bridge would have created a barrier to the passage of marine animals across the strait. In the end, no bridge has been built yet across the Strait of Messina, to this date anyway, for reasons that have nothing to do with the concern for impeding the cetaceans' passage. On that account, however, opponents of the bridge were dead wrong. Dolphins, whales, and a multitude of other marine beings swim under bridges without thinking twice, whenever they feel like doing it, and there is ample documentation of them doing so in famous locations, such as across the Bosphorus connecting the

Black Sea with the Marmara Sea in Türkiye, or across the Golden Gate in California. As far as the Rion-Antirion bridge is concerned, in nine different instances—and these are only those we know about—bottlenose dolphins from further to the north, such as the Gulf of Ambracia and the Inner Ionian Sea, were resighted inside the Gulf of Corinth. Even a mighty fin whale, the largest of all Mediterranean animals, was reported to pay a visit to the Gulf of Corinth in 2011, having transited under the bridge apparently without significant apprehension.

The Corinthian Gulf is a special inner sea: a mini-Mediterranean nested within the bigger Mediterranean, surrounded by land on all sides except for the very narrow entrance to the west where I was coming in from. Despite its small size—being only 130 km long and slightly more than 30 km wide at its widest point, with a surface of exactly one-thousandth of that of the Mediterranean—the gulf reaches a respectable depth of 935 m. Its deep waters create a peculiar abyssal mini-ecosystem, in particular, hosting an important population of mid-water cephalopods such as long-armed squids and umbrella squids.[16] The squids, in turn, are food to predators such as blue sharks and pelagic dolphins, which would normally be typical of the open sea and, therefore, absent from this area.

As I progressed into the Gulf in the warm afternoon light, a decent breeze flowed down from the Peloponnese valleys, helping to push *Pontoporia* forward under sail, significantly increasing her speed, and helpfully giving some welcome rest to the boat's engine and my sense of hearing. The sun was already getting low in my wake when a school of striped dolphins[17] came to play in front of the bow, brightening a day in which I had seen lots of boats and people but very little wildlife.

---

[16] Respectively, *Chiroteuthis veranyi* and *Histioteuthis bonnellii*.

[17] The striped dolphin, a cosmopolitan species like the common and bottlenose dolphins, was described in 1833 by the German zoologist Franz Julius Meyen as *Delphinus coeruleoalbus*. The species was subsequently classified as *Stenella coeruleoalba*. In the Mediterranean Sea its status has been recently assessed as "Least Concern" (in press). The Gulf of Corinth subpopulation is proposed as Vulnerable (Bearzi G., Bonizzoni S., Santostasi N.L., submitted).

Fig. 4.2 A fast-porpoising striped dolphin displays a pattern of stripes along its flank, which are diagnostic of the species and justify its common name (Illustration by Massimo Demma)

Striped dolphins are very common cetaceans in all oceans and most seas and are the most common dolphins throughout the Mediterranean Sea, from Gibraltar to Cyprus. They are so elegant—with their dark back, white belly, and flanks decorated by dark stripes and shades of grey—that having them around is always a joy for the eyes and the heart, no matter how common an event is meeting a school of them in the open sea (Fig. 4.2).

However, in the eastern portion of the Gulf of Corinth, striped dolphins are special because they are an isolated population of about 1300 individuals, genetically distinct from the rest of the world's striped dolphins.[18] This number is small as far as dolphin populations are normally concerned. However, it is still remarkable considering the gulf's reduced size. It results in a striped dolphin density in the gulf which is almost three times higher than that in the average Mediterranean.

The dolphins of the Gulf of Corinth—also identified as an IMMA for both striped and common dolphins[19]—have recently been the subject of detailed investigations by various colleagues, including, at different moments, Alexandros Frantzis, Giovanni Bearzi, and Silvia Bonizzoni. Surveys by these researchers have revealed that striped dolphins are not alone in the gulf. Bottlenose dolphins have carved for themselves a niche in the somewhat wider shelf and bays along the northern shore, and are the most "casual" cetacean inhabitants of the gulf because they are known to move in and out; at least, some of them do. Two other species, common dolphins and Risso's dolphins,[20]

[18] Bearzi G., Bonizzoni S., Santostasi N.L., Furey N.B., Eddy L., Valavanis V.D., Gimenez O. 2016. Dolphins in a scaled-down Mediterranean: the Gulf of Corinth's odontocetes. In: G. Notarbartolo di Sciara, M. Podestà, B.E. Curry (Editors), Mediterranean marine mammal ecology and conservation. Advances in Marine Biology 75:297–332. https://doi.org/10.1016/bs.amb.2016.07.003.

[19] Gulf of Corinth IMMA: https://tinyurl.com/3akrvfn9.

[20] Risso's dolphin, *Grampus griseus*. I will meet them later in *Pontoporia's* journey, when approaching the Strait of Sicily.

have instead been observed intermingling with striped dolphins in the deeper waters of the eastern end of the gulf. Risso's dolphins are an exception, possibly stray individuals who have arrived in the gulf by chance and decided that it was cool to remain. Years ago, two of them were always associated with striped dolphins; now, only one seems to be left. Common dolphins are more numerous but still very rare: estimates based on photo-identification indicate that, at this moment, there are only 22 of them, their survival in the area hanging from a thread. Common dolphins are invariably found in mixed groups with striped dolphins, and they seem to be doing so well in their company that the two species even interbreed, creating hybrid forms which are apparently fertile.[21] Give them enough time, and we might one day be able to admire a new species in this area: the unique Corinthian dolphin.

That afternoon, the dolphins kept avidly riding the bow wave created by *Pontoporia*, which silently glided forward under sail towards Corinth, my destination of the day. It was getting dark as the evening approached, and I was not sure when the dolphins finally abandoned the moving playground constituted by the microcosm surrounding my boat and resumed their normal delphinine businesses.

There was still a residual glimmer of the day when the lights of the city of Corinth appeared in the distance against the darkening sky, where a quarter moon was starting to shine. As I approached the shore at the end of the Gulf, it was hard not to get the impression that I would end my trip in a cul-de-sac. But there was a way out, of course. The Corinth Canal, dug towards the end of the nineteenth century across the isthmus to create a passage for ships between Ionian and Aegean, was there in front of me hidden like a crack in the solid coastline, and tomorrow will allow me to continue into a new sea.

As I approached the port of Corinth, the contrast was noticeable between the silence and beauty of the natural landscapes I had been immersed in since leaving Otranto and the smell and the noise surrounding this commercial harbour. I was getting too close to civilisation for comfort, with the large city in front of me with its lights, highways, and train tracks. But my escape route through the Canal was there, and tomorrow I was going to use it and be gone again in a novel marine wilderness.

I headed towards the marina nested within Corinth's wider port, only about a nautical mile from the entrance to the Canal. The site was somewhat noisy and unpleasant with the pervading smell of the exhaust fumes of hundreds of trucks and buses roaring about in the surrounding roads. But it was good to rest after that long day.

---

[21] Antoniou A., Frantzis A., Alexiadou P., Paschou N. 2018. Evidence of introgressive hybridization between *Stenella coeruleoalba* and *Delphinus delphis* in the Greek Seas. Molecular Phylogenetics and Evolution 129:325–337. https://doi.org/10.1016/j.ympev.2018.09.007.

5

Aegean Sea: From Corinth to Rhodes

Illustration by Massimo Demma

Day 10: Corinth, Greece

Eos was painting the sky with her rosy fingers, to borrow one of the most endearing of Homer's metaphors, when I left the dock at Corinth's marina and eagerly pointed *Pontoporia*'s bow towards the entrance of the Canal. I

© The Author(s), under exclusive license to Springer Nature Switzerland AG 2024
G. Notarbartolo di Sciara, *Sailing Across a Wounded Sea*,
https://doi.org/10.1007/978-3-031-54597-9_5

approached the crossing with anticipation and mixed feelings. The Canal is a narrow waterway cut across the isthmus connecting the Greek northern mainland to the Peloponnese peninsula. I felt deep admiration for the labourers of the past, who had made the passage possible by digging a canal across several kilometres of solid land in the nineteenth century, when modern earth-moving machines and similar contraptions were still beyond anyone's wildest dreams.

Deciding to go this way felt almost like being a Fitzcarraldo forcing my boat from one sea to another through land, somewhat unnatural and a bit of a cop-out, largely to avoid the trouble of circumnavigating the Peloponnese, and to reach the Aegean sooner. But, in fact, Werner Herzog's story about the Amazonian steamship transported across a mountain to another body of water, in order to avoid more dire passages, was predated by a little-known page of ancient history that unfolded exactly here. To enable the movement of boats across the Isthmus of Corinth, thereby avoiding the lengthy and perilous circumnavigation of the Peloponnese when transiting between the Ionian and the Aegean seas, the clever Greeks had built a trackway over the isthmus paved with hard limestone, called the *Diolkos*, where boats could be pulled across land. The trackway was in operation for no less than seven centuries, starting at about 600 BCE.

The prospect of crossing the Canal presented itself as an attractive adventure but ripe with unknowns. Queues of colossal ships dwarfing my frail vessel whilst waiting for their turn to pass; endless waiting time at anchor in precarious conditions; exorbitant crossing fees; and other such challenges had haunted my sleep the night before. As it turned out, none of that materialised. The Corinth Canal is a rather sleepy waterway nowadays. Too narrow to allow the passage of most modern-day ships, today the Canal sees traffic that is much more limited than it used to be. When I arrived at the Canal's entrance I was alone, and I was admitted right away by a portly attendant housed in a small, shabby shack. Moreover, the fee I was charged at the eastern exit of the Canal seemed a small price to pay indeed, not only for the convenience of a significantly shortened route, but also for the emotions that the passage evoked.

As *Pontoporia* proceeded along the short waterway—the Canal is a mere 3.4 miles in length—the banks on both sides were initially low and sloping up only slightly from sea level into the surrounding fields, giving the bizarre sensation of sailing across an olive grove. Despite the early hour, the cicadas were already vigorously stridulating from their arboreal habitats, imbibing my soundscape with their chorus, a typically rural symphony bizarrely out of place when heard from the cockpit of a moving boat. According to Socrates in Plato's *Phaedrus*, cicadas were derived from ancient singers who were so enthralled by their music that they forgot to eat and drink, and therefore died without noticing. Moved by such a sad fate, the Muses gave their hemipteran descendants the gift of singing incessantly without food, from birth to death. This is not really true, as we now know that adult cicadas, seemingly solely intent on emitting the trademark call of hot Mediterranean summers, survive by unobtrusively sucking the sap from the branches where they perch. But it is a charming tale nevertheless, like so many of the myths that grew out of the cultural broth of the Aegean Sea, in which I was about to immerse myself.

The land bridge bisected by the Canal gradually rose in altitude eastwards, and soon I found myself navigating in a waterway deeply cut between limestone walls, giving me the false impression of proceeding along an increasingly narrow trench. High above, from a bridge across the Canal, tiny human figures were waving at me. Then, suddenly, the limestone walls were gone, making room for the widest and bluest of skies, delivering me again into a more familiar marine landscape. *Pontoporia* had entered the Aegean Sea.

Visiting the Aegean has always been a momentous event for me because this sea contains endless, admixed natural and cultural wonders. The goal of my Mediterranean wanderings aboard *Pontoporia* was, admittedly, in quest of natural as opposed to historical discoveries. However, cultural background is an important filter, accepted willingly or not, through which our eyes appraise the surrounding world. For those of us from the Western world, so much of our cultural fabric originated in this archipelagic playground where the stories of gods and heroes, scholars and scientists, poets, adventurers, pirates, and seafarers are woven together in a seamless continuum, from myth to history and from philosophy to science. Indeed, the Aegean Sea is the primaeval soup bathing the lands where, to a large extent, my civilisation was incubated. Again, as mentioned when I first entered Greece days before, when arriving here, I felt this ancestral, almost hard-wired sense of coming home.

At times, however, the feeling of connectedness with the past seems stronger in the imagination than it is in reality. Over the millennia, time here has managed to pile so many layers of narratives, one on top of another, that the physical and societal landscape has changed beyond recognition even between now and one hundred, let alone thousands, of years ago. Factual memories of everyday happenings in the ancient past are all but lost with minor exceptions, and most of what remains are the stories told by the few stone artefacts that have managed to defy time. When walking in places that I know were built, inhabited, or worshipped by the ancients, for example in the Athens agora, I find myself struggling to picture in my mind what these places were like when they were inhabited by the people who built them. Most of the time, my imagination fails amidst the tourist crowds. But some places still exist, protected by special circumstances from noisy crowds and the trash they bring along, that have preserved their ability to transmit their situational personality to the discerning bystander, allowing the illusion of time travel to the receptive mind and blissfully conveying a sense of continuity between past and present. I felt the magic of the Aegean Sea, knowing it to be ripe with such places.

Sandwiched between the Balkan and Anatolian peninsulas, the Aegean Sea is unique in the Mediterranean region because of its relatively shallow depth—on average less than 1000 m, with a few isolated deeper portions, compared with the Mediterranean mean of 1500 m—and its more than 7500 islands, islets and rocks, only a small portion of which are inhabited. Each of these islands comprises its own microcosm, with its distinctive appearance, geology, naturalness, history, and humanity: in a word, personality.

This abundance and diversity of lands and islands, ranging in size from the few tens of square kilometres of any of the Little Cyclades to the more than 8000 km^2 of Crete, was a source of wonder and pride for the ancients who considered the Aegean to be the "prince of all seas", which is the true meaning of the word "archipelago". It was only later, in a bizarre example of semantic metamorphosis, that the term archipelago lost its original meaning and came to designate a discrete cluster of islands. In fact, amongst the multitude of islands in the Aegean, several are organised into separate archipelagos, such as the Cyclades in the south, the Sporades in the northwest, and the Dodecanese in the east, right alongside the coast of Asia Minor.

This swarm of islands encouraged more than the flourishing of a multitude of human settlements and cultures along the Aegean shores. The morphological complexity of the interface between water and land, regularly subjected to the region's forceful, often violent atmospheric and oceanographic manifestations, is also the source of a variety of environments and habitats, able in turn to host and protect from human insult the special assemblages of living organisms that I was coming to visit. Sporadic upheavals caused by a restless Earth's crust, peppered with volcanoes and shaken by the occasional earthquake, have been a force of change in the area since the dawn of times, often accompanied by raging windstorms and pounding turbulent waters, considered by the ancients the remit, respectively, of Aeolus and Poseidon.

With the highest expectations fuelled by these thoughts, I set *Pontoporia*'s course to the southeast from the outlet of the Canal, headed across the Saronic Gulf. One of the many large indentations along the eastern coast of mainland Greece, the Saronic Gulf is formed by the Attica Peninsula to the north and the Peloponnese Argolis Peninsula to the south. My plans for the day were to sail into the heart of the Aegean about 90 nautical miles to the east, to the middle of the Cyclades, and to end my day's travel in the lee of the uninhabited island of Gyaros, believed to be one of the main Mediterranean strongholds of monk seals.

I knew that by venturing into the Aegean Sea on a small vessel in summer I would have to be prepared for tough going. Summer is the season in which the region is dominated by the Meltemi, also known as Etesian winds. Meltemi—presumably a lexical loan from the Italian "mal tempo" (bad weather)—is a meteorological feature endemic to the Aegean, blowing from June to September and produced by the interaction between two distinct atmospheric features that build up during the hot season in a roughly static fashion: a high-pressure centre over the Balkans and a low-pressure area over the Anatolian peninsula. In a combined conveyor belt process, by rotating in opposite directions—clockwise the first, counter-clockwise the second—the two systems draw in the cooler, dryer air from northeastern Europe and the Black Sea and project it down, more or less violently (depending on the pressure difference between the two systems), across the Aegean Sea.

A strong Meltemi, typically 7–8 Beaufort[1] or more, can be dangerous and likely to affect the operation of even the largest ferries. By contrast, when it is weaker, a "Meltemaki" (small meltemi), as the Greeks call it and thankfully its most common form, creates quite pleasant conditions with clear, fine weather, thereby blessing the area with a cool, dry environment, in stark contrast to the sweltering heat experienced elsewhere during the Southern European summer. Ultimately, apart from moments of extremes, the Meltemi is a muscular but welcoming presence for the Aegean visitor that makes this part of the Mediterranean uniquely pleasant. With the Meltemi "on", the sunsets and dawns acquire a typical pinkish haze, the nights are cool and conducive to fine sleep, and the wind's presence is marked by the low, constant hum of rushing air, forceful but devoid of the nastiness that is typical of storms. By contrast, with the Meltemi "off"—as occurs occasionally for a few days between windy spells—the air becomes hot and humid, leaving one counting the hours for the Meltemi to return. Most importantly, the sailor who endures the whipping of the Meltemi at least knows exactly what to expect from the god of Mediterranean winds and can rest assured that, despite some tough going, the situation will remain, to some extent, manageable. Unfortunately, this is not the case with the tropical-like cyclones once unknown in the Mediterranean, but that are happening with growing frequency and destructiveness of late,

---

[1] The Beaufort Scale provides a means of describing sea conditions resulting from wind speed, and ranges from 0 (no wind, calm water) to 12 (hurricane conditions): https://tinyurl.com/s72uvxda.

fuelled by the greater temperature differences between sea and atmosphere resulting from global warming.

Navigating across the Saronic Gulf turned out to be relatively tranquil because of the shelter from the Meltemi provided by the Attica peninsula. Still, I knew that conditions would change once I left the lee provided by the land beyond Cape Sounion, the peninsula's southeastern tip. A moderate breeze from the northeast coming abeam moved *Pontoporia* smoothly and briskly. My main concern as I was cruising across the strait between the islands of Aegina and Salamina was not the weather but maritime traffic. Vessels of all sizes and shapes were everywhere: ferries exiting and entering the port of Piraeus to the north; dozens of tankers anchored in the lee of the island of Salamina, waiting to offload their filthy brew; pleasure crafts irradiating from the coast to bring tourists to their dream destinations; and even a regatta of small white sails to the east of Piraeus, pleasantly dotting the gulf's blue expanse.

The names of these places were bringing back grade-school memories of the feats of the Athenian admiral Themistocles, who in 480 BCE managed to defeat the powerful fleet of the Persian king Xerxes with his smaller but nimbler vessels right there, in the narrows between Salamina and the mainland. The contrast could not have been starker between my mental image of the fleet of tiny Athenian vessels clashing with the ships of the mighty invader 2500 years ago, and the current aspect of the place, with the coast in front of Salamina mangled by a skyline of oil refineries and giant storage tanks of crude, and the multitude of tankers moored nearby in patient attendance.

I sailed rapidly across the gulf, nearing the coast of Attica to port, as I headed towards Cape Sounion. The heavily developed coastline, with the port of Piraeus and Athens' metropolitan sprawl spreading inland as far as the eye could see, was progressively yielding to sparser urbanisation as I gained distance from the city, and the vessel traffic lessened as I was leaving the main shipping lanes. I felt like I was reappropriating my space in the sea, and it was a good feeling to leave the most disturbing symptoms of human-caused degradation in my wake.

And then, in the warm glow of the afternoon, the ruins of Poseidon's temple atop Cape Sounion came into view. I found myself marvelling at the emotion produced by a few ruined stone columns rising from that arid headland, a sight so powerful, despite having seen it many times before, that it trespasses from the conceptual into the instinctive and from mind to heart. This was finally a place where I could contemplate an intriguing vestige of the past, surrounded by a landscape where not much existed to emphasise the disconnection between past and present. Here, it was possible to share with a shiver the sentiment of the ancient Athenian seafarer bidding goodbye to the outermost signpost of his homeland with a prayer to Poseidon to let him return safely to the comfort of his home walls and affections. Unlike today, places were dangerous and travel uncertain in those times, and many seafarers did not survive to return home.

Historian Adrienne Mayor relates a curious anecdote of Sir Arthur Conan Doyle, who told of having observed a sea monster in the waters off Cape Sounion when standing on the steamer's deck as it was rounding the cape. Whilst gazing at the ancient ruins, his eyes were distracted by the appearance in the waves of a "curious creature" slightly longer than a metre that "had a long neck and large flippers. I believe, as did my wife, that it was a young plesiosaurus".[2] The Conan Doyles did not have more than a fleeting view of the mysterious animal, which, in retrospect, was likely a young monk seal. Plesiosaurs were marine reptiles that became extinct 66 million years ago.

My scholarly ruminations ended abruptly beyond Cape Sounion when the first fierce gust of a Meltemi no longer restrained by the shelter of the Attican land caused what seemed to be a bucketful of cool seawater to land on my face. I was not totally unprepared, however: feeling *Pontoporia's* heel over until she took water from the starboard gunwale was an effective wake-up call. I reduced sail, and things seemed right again, with the boat now galloping over the waves like a bronco, with a favourable strong wind abeam.

Crossing the open strait between the Greek mainland and Kea, the first of the Cyclades islands to encounter, was a bit of a tense ride nonetheless, with the wind's muscular energy causing all of *Pontoporia's* components to creak in protest. As the sunset approached and the shadows lengthened, I could not avoid feeling an ancestral sense of unease at the prospect of being alone on a small boat in rough seas with night just around the corner. But everything went well, and soon I was in Kea's lee. An effluvium of aromas from the sunbaked scrub upwind, wafting on me, had a strong reassuring effect, whilst the drying sea spray on my face had given me a pleasantly stiff, salt-encrusted beard, as I sailed across the flat waters in the island's shelter. Once I rounded Kea's south cape and exited the lee, the island of Gyaros, my destination for the day, came into view and approached fast. Rest was in sight, and the thought was welcome.

There is a bay on the southeast coast of Gyaros that is well-sheltered from northern weather, and that is where I dropped anchor, just as dusk was yielding to dark. Powerful gusts raging down from the island's mountain caused *Pontoporia* and her mast to vibrate at times, but I had had the good sense to

---

[2] Sir Arthur Conan Doyle and the Sea Monster: https://tinyurl.com/5dk76j3u.

endow *Pontoporia* with a large anchor with a long, heavy chain, and the bay's sandy bottom offered a solid, safe anchorage. It was so sweet, after the long day, to place my body horizontally in the cool, dry sheets and plunge into a long, dreamless sleep.

Day 11: Gyaros, Greece

As daylight replaced the darkness, and the Moka coffeemaker greeted me for a new day with its comforting, gurgling song and fragrance, the details of my surroundings came into view, revealing the island's barren, unforgiving landscape. The bay was framed by a rocky coastline scattered with low bushes of thorny burnet and thyme, sloping up to the top of Gyaros, almost 500 m above me, without a single tree in sight. To my left, a huge, intimidating, semi-ruined building was perched with its red walls on a flat headland jutting out into the water, frowning at me with its series of frameless windows, resembling the empty eye sockets of a multi-eyed monster. The sight of the decaying building looming over the bay, surrounded by such a stern landscape, gave me a distinct sense of unease. Not without reason.

Gyaros was a place of suffering, forced labour, and torture for millennia until only a few decades ago, and as such, it is known to older Greeks to this day. Used as a jail and place of exile since Roman times, Gyaros gained its worst reputation between 1947 and 1952 when political opponents—up to 20,000 of them—were imprisoned there by the Greek military junta. As memories crumble under the assault of time, not unlike the building where those events unfolded, there was a sense that they nevertheless still lingered and stood, just like the prison's walls: a reminder of that ugly past, not so distant, and that could recur again in the blink of an eye.

But I had come to Gyaros for something much more cheerful. The isolation from human encroachment, first as a penal colony, and then, following the fall of the junta, as a militarised area devoid of human inhabitants, had unintentionally transformed Gyaros into a haven for a variety of endangered Mediterranean wildlife—barring the occasional bouts of shellfire when the cliffs were used as a practice target during naval exercises, an insane practice thankfully discontinued after the year 2000. Lack of public access minimised disturbances, allowing the local fauna to reassemble and flourish.

Fig. 5.1 Wary of the proximity of the researcher behind her, a Mediterranean monk seal female swims away to maintain a safe distance but is also curious about the intruder (Illustration by Massimo Demma)

Today the island hosts one of the largest breeding colonies of Yelkouan shearwaters,[3] threatened marine birds endemic to the Mediterranean, and its steep western cliffs are home to a sizeable community of Eleonora's falcons.[4] Named in honour of the fourteenth-century queen of Sardinia, Eleanor of Arborea, these beautiful falcons are unusual birds of prey, with their habit of living in colonies and of migrating annually between their summer Mediterranean nesting sites and their wintering grounds in Madagascar. I never tire of watching their collective aerial acrobatics. European shags, common kestrels, common buzzards, and even one pair of stately Bonelli's eagles[5] further enrich the notable avian fauna of Gyaros. Even more interesting to me than the birds, however, was the recent discovery that Gyaros might be one of the world's most important sites for Mediterranean monk seals. Most of the southern Aegean, including the waters around Cyclades and the Dodecanese islands, was identified as an IMMA for monk seals.[6] Still, Gyaros sits right in the middle of it and is apparently prominent in its conservation relevance (Fig. 5.1).

A pause is in order here to explain the full significance of this discovery. Of the pinniped group, there are about twenty species of phocid (true or earless) seals widely distributed across the world's oceans and inland waters (the other members of the broader group of pinnipeds are the walrus and the eared seals, or sea lion-type species). Most phocid seals are attractive and cute, especially when young. Except for the large and somewhat intimidating adult males,

---

[3] The Yelkouan shearwater, *Puffinus yelkouan*, is currently assessed as "Vulnerable": https://tinyurl.com/4df33wb2.

[4] The Eleonora's falcon, *Falco eleonorae*, is assessed as "Least Concern": https://tinyurl.com/2janb48u.

[5] The Mediterranean population of the Bonelli's eagle, *Aquila fasciata*, is listed as "Least Concern" but decreasing: https://tinyurl.com/mfjyrnm7.

[6] Central Aegean IMMA: https://tinyurl.com/ycx8y9nk.

Mediterranean monk seals can also be seen as attractive and cute, but that is not the reason they are special. Neither are they special because, unlike most phocids, they tend to live in warmer waters.

As I noted a few days ago when visiting the inner Ionian Sea, monk seals are special because they can be considered symbols of the value of the Mediterranean marine habitat, their diverse and rich habitat disappearing under the effects of human overexploitation, encroachment, and mismanagement, leading to an unjust outcome in which both parties, seals and humans alike, are left to suffer. Most importantly, their hoped-for recovery could convey an even stronger message: restoring the broader Mediterranean marine and coastal environment is possible if we try hard enough.

There once were three monk seal species in the world's oceans. Like its Hawaiian cousin,[7] now numbering fewer than 1000 individuals, the Mediterranean monk seal is one of the most endangered of all marine mammals, its numbers no greater than a few hundred individuals. However, an even worse fate befell the third species, the Caribbean monk seal.[8] First recorded by Western observers during the second voyage of Christopher Columbus, Caribbean monk seals are now extinct, perhaps the least acknowledged of the many casualties inflicted on the inhabitants of the Americas, human and non-human alike, by the European invaders.

Today Mediterranean monk seals are so rare and skittish that coming across one is like winning the lottery. That is why I was so enthralled a few days earlier by my experience with them in the Ionian Sea, and that is why I had planned the visit to Gyaros.

The discovery that Gyaros might host an exceptional aggregation of monk seals was made in the early 2000s, when the military permitted civilian researchers to explore eight large caves carved in the island's steepest cliffs. Some of these caves were found to be in use by the seals as quiet, secluded locations for birthing their pups. The number of newborn seals observed during the breeding season (late summer to autumn) led the researchers to estimate the Gyaros colony at about sixty individuals. This is a derived figure obtained by counting the pups found in one season within the complex of marine caves perforating the island's coastline, and multiplying the count by an empirically-derived constant. So, the number is open to debate, given that, in fact, 60 adult seals had never been seen nor counted around the island.

---

[7] The Hawaiian monk seal, *Neomonachus schauinslandi*, is classified as "Endangered": https://tinyurl.com/4d4mhh9p.

[8] The Caribbean monk seal, *Neomonachus tropicalis*, is now extinct: https://tinyurl.com/ykja48t6.

However, if true, this estimate would make Gyaros the most important monk seal aggregation site in the whole of the Mediterranean.[9]

Therefore, I planned a visit to the coastal cliffs of Gyaros, with the expectation of chancing upon some of the elusive pinnipeds, and with the hope of being as lucky as I had been in the Ionian Sea.

My plan was to circumnavigate the island with an exceptional guide, Spyros Kotomatas, a colleague and long-time friend, in charge of a project funded by the European Commission to protect Gyaros and its monk seals. Spyros joined me that morning in Gyaros with the fast vessel from his project base in nearby Syros, about 30 min away. A population ecologist by training, with a doctorate from the University of Illinois, Spyros has dedicated the greater part of his professional life to the conservation of Mediterranean monk seals, first as the managing director of the Hellenic Society for the Study and Protection of the Monk Seal (a.k.a. "MOm"), and later as a senior marine conservation officer with World Wildlife Fund (WWF)—Greece. In my losing struggle to perceive continuity between classical and modern Greece, I consider Spyros to be an exception. I see him perfectly impersonating the archetype of the thinker sitting on a marble bench in the Athens agora a few thousand years ago, perhaps arguing with Democritus about the fundamentals of life on Earth—that is, assuming that the philosophers of the time were chain smokers, and wore a ponytail as he does.

One day in April, several years earlier, Spyros and I were sipping *metrio* coffee under the most classic vine-covered pergolas in Monastiraki (a district of Athens), discussing ways to address the many challenges of protecting Mediterranean monk seals. I had been careful to sit upwind to avoid being engulfed in the cloud of smoke emerging from his constantly lit cigarette. At the time, Prince Albert II of Monaco, one of the few European heads of state whom I know to be genuinely and actively concerned with the conservation of the Mediterranean environment and aware of the symbolic value of monk seal conservation, had publicly announced his commitment to the cause of the beleaguered pinniped. His foundation asked me to make recommendations concerning sensible activities to stem the decline and promote the recovery of the species.

Spyros and I were well aware of some outlandish proposals received by the Prince's foundation, such as capturing monk seals in Mauritania or the eastern Mediterranean and releasing them in the western part of the region to help

---

[9] Dendrinos P., Karamanlidis A.A., Kotomatas S., Paravas V., Adamantopoulou S. 2008. Report of a new Mediterranean monk seal (*Monachus monachus*) breeding colony in the Aegean Sea, Greece. Aquatic Mammals 34(3):355–361. DOI 10.1578/AM.34.3.2008.355.

recolonise that portion of the sea from which they had been so thoroughly and effectively extirpated. We knew that similar reckless experiments with relocating endangered species, having considerable mediatic potential but groundless in terms of conservation practice, might instead have helped to push off the cliff a species already on the brink of extinction. We had argued that, instead, by far the best chance of helping monk seals regain their former Mediterranean haunts, including the western portion, was to secure their foothold in Greece, where small but viable breeding communities were still holding fast. This strategy would lead the seals to eventually expand westward on their own and recolonise their former habitat. The importance of Gyaros had recently been discovered by MOm's researchers, but the site was devoid of protection for the seals other than its inaccessible status as a military area, which could vanish at any moment by the stroke of an institutional pen. Protecting monk seals in Gyaros was, therefore, a matter of great urgency.

Amongst the main actors concerned with monk seal conservation in Greece at that time, consensus was rapidly developing that betting on Gyaros was a winning card. Thus, a 4-year project was hatched, which was eventually supported in large part by the European Commission with the participation of Prince Albert II of Monaco Foundation and a few other smaller donors. The project was articulated into a complex set of complementary actions, which included commitments to protect the site by the relevant institutions at all levels, from local to national to European. The project also ensured the continuity of systematic conservation measures, such as monitoring and guarding. Most importantly, the project addressed the social and economic aspects of protecting Gyaros by facilitating participation in decision-making through a consortium of key representatives of the local communities. The lead organisation of the project, named "Cyclades LIFE", was WWF Greece, and Spyros was its coordinator.

I jumped from *Pontoporia* onto Spyros' nimble inflatable to circumnavigate the island and venture into prime monk seal habitat. Luckily Lord Meltemi had partly calmed down from the previous day, permitting the safe, albeit slightly bumpy, exploration of the windward side of the island, the side preferred by the seals. The coast there is made up of steep, dark, formidable-looking cliffs interspersed by small black pebble beaches and perforated by marine caves: pure perfection as far as monk seal habitat is concerned.

The simple fact of entering the seals' habitat and being part of the scenery was a reward in itself. The small boat moved over the deep blue waters, lifted up and down by the swell as Spiros gingerly negotiated the rocks under the shade of the vertical cliffs blocking half the sky. All my senses were bathed in the forceful manifestations of that primaeval landscape, devoid of any unwelcome sign of human presence. The salty smell of seaweed pleased the nostrils, whilst the thumping sound of the surging seas against the rocks was punctuated by the cries of a multitude of sea birds, some perched just above the spray zone and some circling overhead. I was amidst a display of natural power, bathed in an unspoiled pureness that went through the bones. Had that bird clatter been emitted instead by a community of pterosaurs, I would not have been too surprised. Nor would have Sir Conan Doyle.

Unfortunately, no seals were to grace us with their appearance that day. I could not help noticing how different this experience was from the idyllic scene in the inner Ionian Sea archipelago a handful of days before. There, in a setting of dream-like tranquillity, with *Pontoporia* anchored in the calmest of waters of a small, sheltered bay of Formikoula, I had been approached by seals happily minding their own business within a stone's throw away, as if I did not exist.

My disappointment at not meeting monk seals in Gyaros was mild, however. I had had the privilege of being admitted to their Gyaros residence and felt in awe of its majestic aura. If the seals happened to be somewhere else on that day, that was unfortunate, but I could hope to meet them some other time. It would be foolish to consider the open sea like an aquarium: you cannot ask for your money back if you have not seen what you wished to see. Besides, I knew the seals were there. The rapscallions were likely hiding from us just under the surface or behind a rock, sniggering under their moustaches. Hiding from humans, after all, is what enabled monk seals to survive until this day. Perhaps one of the peculiarities of the Ionian Sea is that there, unlike in the Aegean where human persecution still occurs, the seals might have become more relaxed in the presence of humans, and dispensing with an absolute need to remain hidden.

I was curious about Spyros' opinion of the project he led, which was just about to end.

"Has the monk seals' future in Gyaros been made more secure through the project?" I wondered.

"Certainly yes," said Spyros, "and not only in Gyaros, but also within a much wider radius extending far beyond the island, because if Gyaros continues to be a safe haven for the seals, their pups will eventually irradiate into the surroundings as they mature."

"What was the Greek government's role in the success of the project?" I asked.

"The government has formally extended legal protection over Gyaros as part of its contribution to Cyclades LIFE", Spyros told me. "The protection includes a 3-nautical mile zone surrounding the island where fishing is prohibited, superimposing a layer of national legislation with the formal establishment of an MPA on top of the pre-existing denomination of Natura 2000 site under European law".

"What protective infrastructure will the project leave in place once it's finished?" I continued with my questioning.

"The project has given the Gyaros MPA key enforcement tools installed on the very top of the island," said Spyros. "These include a radar and a powerful high-resolution, remotely operated video camera, both of which are now used by the Gyaros MPA Management Agency and the Coast Guard in Syros to monitor compliance of the no-fishing zone".

In fact, looking up, I could see the small white building on top of the island where the radar and the video camera are housed. If I could see the camera, the camera could see me, which was the goal.

"If the radar detects a boat with suspicious behaviour entering the no-take zone, where fishing is banned, at any time of day or night, the camera can be used to identify the boat, and then the Coast Guard can intervene", said Spyros. "The routes and speeds of boats when fishing clearly give away what they are doing".

Spyros then summarised for me his work and the knowledge gained over the 4 years of the project. "We mapped marine habitats down to a depth of 160 meters to identify the habitats protected by law, such as seagrass and coralligenous bottoms, where fishing cannot occur; we monitored the ecology, behaviour and breeding of the monk seals on the island; and we completed a census of the marine bird colonies. Detailed socioeconomic analyses of the human communities from the nearby inhabited islands of Syros and Andros, which gravitate around Gyaros, showed us how important the site is for local people and helped us find ways to gain local acceptance of a more conservation-minded mentality".

One thing was clear from what Spyros was telling me: monk seals can be protected only by addressing the human part of the human/seal relationship. There is nothing inherently problematic with the ability of monk seals to survive; the species would be flourishing in the absence of human agency. The challenge consists solely of changing human attitudes, of convincing the local people—fisherfolk in particular—to accept the seals, to learn how to coexist with them, and to tolerate the cost of damaged nets and catch losses that such coexistence often entails.

"However, isn't it difficult, and a bit unfair, to expect these economically marginal communities to bear the cost of dealing with a healthy population of seals merrily competing with them for the few remaining fish?" I asked.

"Bearing the cost of economic damages inflicted by a handful of seals on fisheries might involve hardship for the individual fisher but is obviously not a major issue for the broader society and the State, which should at least partially address the issue through various compensatory mechanisms", said Spyros. He then explained how the project helped to establish a consortium of stakeholders, including representatives of local administrations, fishers, the tourism industry and the relevant authorities that have been tasked with discussing the benefits and costs of conserving the area and its fauna, monk seals included of course, and creating a mechanism for sharing this debate with the wider local community.

"I think establishing the local consortium is really the most important and unique legacy of the project", said Spyros. "I hope it sets an example for many other similar situations in the Mediterranean, in Greece and elsewhere, where the well-being of the local communities is made to coexist with the well-being of their marine environment".

"How did the locals react to your efforts to involve them in the conservation process?"

"We were pleasantly surprised by how receptive the local community was", Spyros replied. "Despite our resolve, we had been pessimistic about the outcome of such daring decisions because prior conservation initiatives on other Greek islands were nipped in the bud by strong local aversion to any change. So in the end, we were astonished at how well things went here", Spyros concluded.

Perhaps it will be the destiny of Gyaros, a name still associated with dark memories of human suffering and injustice, to be remembered as a model of reconciliation between Mediterranean islanders and their environment, transitioning from a state of belligerence to one of appreciative coexistence, I mused, as Spyros steered the boat along the island's southern shore.

Nowadays, the socioeconomic status of small Greek island communities is buffeted by a tsunami of societal change, and the entrenched conservativeness of the archaic agricultural and fisheries habits is under assault by changing scenarios, such as income from tourism. The protection of the environmental component of their world gained importance first as a concern and is now seen as an opportunity. Indeed, as I mentioned earlier, the decline of Mediterranean monk seals in their few remaining strongholds—Greece, Cyprus, and Türkiye, Madeira and Mauritania—has now slowed, and it even appears to be cautiously reversing in places. Accordingly, the species status has improved from being considered 'Critically Endangered' to being only 'Vulnerable' in the Red List of Threatened Species compiled by the International Union for the Conservation of Nature.[10]

The causes of this portentous reversal of the species' karma, at least as far as Greece is concerned, are certainly more than one. To begin with, the geography of the Aegean provides ideal habitat and shelter with its myriad of uninhabited islands and islets—largely inaccessible because of the area's strong winds, particularly with the summer Meltemi, thereby providing the seals with a refuge from human encroachment and persecution that they do not have in other Mediterranean locations. Here they have survived in sufficient numbers, despite decades of persecution, to remain a viable population in wait for kinder times to come.

Secondly, a more favourable outlook for monk seals might be coinciding with the progressive decline of small-scale fisheries in Greece. Fish stocks have been severely overexploited through decades of despicable mismanagement and are declining everywhere but especially in the poorly productive waters of the southern Aegean. Making a living as an artisanal fisher has become a marginal feat at best and an impossible one for many. Therefore, the children of fishing families have become increasingly involved in the more lucrative and less arduous tourist trade developing across most of the Aegean islands, with fishing activities limited to the off-season and increasingly acquiring the character of a secondary job or even a hobby.

Meanwhile, a third factor—broader societal change—is bringing about a less hostile attitude towards nature and wildlife in general and is reaching even these small communities subsisting at the periphery of the global commons. This trend, combined with the decreased economic importance of fishing activities, is likely to be at the root of the observed decrease in fishery-derived seal mortality, the prime reason for the seal decline during the last century in the Mediterranean.

So it might be that people saddened by tales of human–wildlife conflicts, occurring globally in the most diverse range of instances pitting carnivores

---

[10] The Mediterranean monk seal, *Monachus monachus*, is now listed as "Vulnerable" after decades of having been listed as "Critically Endangered" first, and subsequently "Endangered": https://tinyurl.com/p2epzfvm.

against herders and large herbivores against farmers, could one day take solace from the Mediterranean monk seal story in Greece. Could this possible happy ending give hope to those worried about the wolves, lions, bears, and elephants succumbing to the war waged against them by humans across the planet?

Soon, I was pleased to see *Pontoporia's* tall mast reappear in the distance at the end of our circumnavigation of Gyaros. It was already early afternoon, and the time had come for me to bid goodbye to my friend Spyros and leave. I intended to put another 30 nautical miles in my wake and drop anchor in Delos for the night. With a lively Meltemi in my favour, covering that distance before sunset was going to be an easy feat.

Several circumstances were making the decision to stop in Delos almost obligatory. First, to have a true Aegean experience, you have to include that island in your itinerary if at all possible. Delos appears no different from any of the myriad small islands scattered across the Aegean Sea with its low, naked, brownish coastline and its diminutive size, less than 5 km along its longest dimension. However, the island's appearance is deceptive. Just to look at it, one would never imagine the amount of cultural intensity that radiated out of this seemingly insignificant rock.

Starting around 900 BCE and for about one thousand years Delos was a major cult centre and the destination of pilgrimages by the devotees of Dionysus and of the twins Artemis and Apollo, who, according to myth, were born there. Delos was also the meeting place of the Delian League, a highly influential coalition of Greek city-states, created to counter the impending dangers of Persian invasions. As a consequence, this history has made Delos one of Greece's major archaeological sites, awarded World Heritage Site status by UNESCO in 1990. There is an amazing concentration of archaeological wonders on the small island, from the several agoras attesting to intense public life; the mosaics indicating residences of the affluent; the temples; the obligate amphitheatre; and a sacred lake where myth has it that the goddess Leto gave birth to her twins, having been impregnated by Zeus. Most magnificent of all, in my opinion, and worth the visit in and of itself, is a terrace where a row of extraordinarily stylised lions carved in snowy-white Parian marble, dedicated to the Sanctuary of Apollo by the islanders of Naxos, stands to guard the place. Today only 5 of the original 12 of these beauties remain on duty. I know one of the missing lions well, as it was stolen by my fellow townsmen in the eighteenth century and now guards the entrance to Venice's Arsenal, the purest Cycladic lines of its body violated by the addition of an ungraceful, disproportionate Baroquish lion's head.

A more personal reason was drawing me to return to Delos. My previous visit was during the long ago summer of 1973, a sombre time when Greece's King had just been deposed, and the infamous regime of the colonels ruled the country by a dictatorship. Greece was mired in political turmoil, and, in retrospect, it probably was not the best time for outsiders like my crew and me to breezily sail around the Aegean for their pleasure. But we had decided to stick to the plans we had hatched the previous winter. Predictably, no other tourists were around.

As we approached the small pier at Delos to dock the boat, the head guard informed us with regret that the archaeological site was closed. Nevertheless, given that the Meltemi was vigorously blowing, in a typical flourish of Mediterranean insouciance of the rules, admixed with the pleasure of being accommodating and showing who was in charge, he said that we were welcome to stay, and even to wander about the island and her ruins.

Appropriating the magic of the entire site for ourselves for a whole day was an extraordinary privilege, indelibly engraved in my memory. In my company was a wonderful young lady with whom I was in love at the time, and for whom fondness mutated into a deep friendship in the decades to follow. Death has now taken her before time could spoil her beauty.

Delos is part of a small cluster of islands and islets, sandwiched between the larger Rineia to the west and the even larger Mykonos to the east. After a joyful ride of a few hours, skirting the upwind coast of Syros with Meltemi on a broad reach and in the relative lee of the island of Tinos, I dropped canvas just before sunset and motored up the narrow channel between Rineia and Delos, made even narrower by the islet of Megali Rematià, and anchored *Pontoporia* just south of the island's only pier. It was too late to visit the ruins again, but I was content with the mere fact of being there, and being able to briefly stroll around the pier in blissful solitude. Retracing a path walked in the past with lost affections under such special circumstances can have the therapeutic effect of coming to terms with reality, accepting the losses entailed, and ultimately finding peace. The brief visit to Delos felt like bringing Eurydice back for a moment from the underworld, but being healed from Orpheus' despair for having lost her.

Half a century after my previous visit, luckily Greek politics had changed, but the winds had not. Meltemi was blowing fiercely now. I dreaded facing the channel between Mykonos and Ikaria with a raging wind, known to raise steep waves, but that was not until the next morning, and I tucked up for the night within the protection of the sacred island. A strong sense of unworldly presence accompanied me through the night, my dreams populated by the friendly ghosts of an increasing cohort of lost affections.

Day 12: Delos, Greece

The day ahead was going to be a mere transfer operation, but first I had to face a rough passage from the Cyclades to the eastern portion of the Aegean, across the notorious unprotected channel between Mykonos and Cape Pappas on the west end of Ikaria. Although the Meltemi was forecasted to be a mere level 6 Beaufort, distinctly less than in the previous days, the crossing was going to be intense, so I had no expectation of marine life encounters over such a rough surface.

For the roughly 10 nautical miles from the southern tip of Delos and along the lee provided by the south coast of Mykonos, I initially sailed fast across a flat sea, swept along by energetic gusts of wind. I had reduced *Pontoporia*'s canvas to the essential sail size to control yawing, and the boat flew ahead like the Scopoli's shearwaters[11] that were dancing around me and skimming the wave crests as if making fun of the violence of the surrounding elements.

As the coast of Mykonos streamed by along the port side, I took comfort from the fact that the distasteful place was at least providing a useful service by creating a lee from Meltemi, smoothing the waves for my passage. Mykonos today is, to me, the negation of all that I love and appreciate about Greece and the Aegean. Through the systematic destruction of the island's once outstanding cultural and natural characteristics, Mykonos has been reduced to a

---

[11] The Scopoli's shearwater, *Calonectris diomedea*, is listed as "Least Concern": https://tinyurl.com/5n7nsr8p.

bubble of nothingness where hordes of party animals converge from the world over to rest their wealthy, or wannabe wealthy, behinds, just like in countless other glitzy cultural junkyards scattered across the world: artificial, appallingly uniform, air-conditioned cocoons complete with chlorinated swimming pools and ear-splitting disco clubs, designed to be carefully eviscerated from whatever is left of the spirit of the places they have contaminated. Simulacra of lunar modules built to allow the survival of a human colony in a desolate moonscape, with the only difference that here the desolation is inside.

Exiting the lee of Mykonos projected me into the unsheltered marine space where the sea metamorphosed from a flat surface, its energy confined to the air, into a melee of brutally corrugated, steep waves, one moment organised in trains as waves are meant to be, and the next in an utterly confused bedlam of whitecaps, with quantities of water hitting against the windward side of *Pontoporia*, and occasionally in my face.

No marine animal had appeared to me in the crossing except for the ubiquitous shearwaters, who deserved all my admiration—and a bit of envy—for their almost cheeky display of nonchalance in that elemental mayhem. Quite unlike the shearwaters, I was in survival mode, keener on taking *Pontoporia* and myself as quickly as possible to safe harbour than on looking for signs of marine life, which would have been hard to spot anyway. In seas like that, I would likely have missed even the largest of whales, unless one were to surface and breach right alongside the boat, something that did not occur that day.

A mere 25 miles were separating me from the shelter of the closest land ahead, the island of Ikaria, so called because it was said that Icarus fell out of the sky into its nearby seas. Ikaria is a long, mountainous, forested island, famous because its inhabitants are graced by one of the world's greatest longevities. In the end, it was not such a long sail to reach Ikaria after all, as *Pontoporia* made good headway with reduced canvas and handled the elements' abuse well. That said, my tension was inevitable, as I knew from experience that all is well in a rough sea … until something breaks.

Thankfully nothing broke, and about three hours later I was again in flat seas downwind from Ikaria. I became more relaxed, as progress was now rapid and *Pontoporia* was sailing well despite occasional strong gusts hurtling down from the island's steep slopes that caused her to broach quite violently. The hours sped by as I sailed first alongside the south coast of Ikaria, then south of the small island group of Fourni, before finally reaching Samos, my destination for the day.

The name Fourni evokes images of piracy because this was one of many Aegean islands serving as a pirate base for centuries. It is difficult to imagine today, when cruising in Aegean waters, and in most Mediterranean waters for

that matter, that in a not-too-distant past—until the mid-nineteenth century—it was rather unwise to sail in the Aegean without the protection of an
armed escort, and without considering the possibility of encountering pirates.
The malign influence of pirates on coastal communities was a constant ingredient of life in the region. Important elements of island societies were shaped
within an ecology of fear, the same way that the presence of lions has shaped
the behaviour of impalas on the Serengeti plains. On most Aegean islands, for
example, the main towns, often fortified with high walls, are located on a hill
at some distance from the coast to give the fishing communities the time to
run uphill from the indefensible harbours and take refuge.

Records of pirate presence in the Aegean date back to classical mythology.
Dolphins, now considered a benevolent, joyful, and human-friendly tribe,
were for the ancients nothing less than pirates who had abducted the god
Dionysus from Ikaria and were transformed into their current shape after having been thrown overboard by the irate god. The pirates who kidnapped Julius
Caesar during one of his Aegean travels were not as lucky. Having been
released from imprisonment in Farmakonisi after the ransom was paid, Caesar
came back with an armed fleet and crucified his captors, keeping a promise
that the hapless pirates had perhaps not taken seriously enough.

The practice of piracy in the Aegean Sea continued across the centuries,
with famous characters engaged in extensive forays giving a hard time to the
everyday exercise of navigation and trade. Hayreddin Barbarossa and Lambros
Catsonis were amongst the most notable of the genre, the former operating
largely on behalf of the Ottoman Sultan at the turn of the sixteenth century,
and the latter, on the opposite side, harassing the Ottoman fleet 200 years
later. These two were the richest and most famous of the lot, and they engaged
in the profession with whole fleets and on behalf of governments; we can read
about their deeds in history books. By contrast, Aegean piracy as a phenomenon mostly concerned diffuse operations conducted by small, unnamed
groups of marine brigands scattered across the region, who engaged in localised actions pitting poor against poor, to the detriment of the local communities. These actions were of limited consequence in terms of scale but had
long-term effects on the social and economic development of Aegean coastal
peoples.

Soon, under the wind's push, Fourni and my reflections on piracy both
faded astern as Kerkis—Samos' imposing mountain peaking at more than
1400 m—rose above me, and I steered for the island's main leeside harbour,
Pythagoras' birthplace. Right across from Pythagorion, as the town is aptly
called, and a mere 3 nautical miles away, lies the coast of Türkiye, with the

Mycale mountains looming over the narrow strait separating Europe from Asia. I had finally reached the definite boundary between East and West.

Day 13: Samos, Greece

The garland of Greek islands in the southeastern part of the Aegean, south of Samos and along the coast of the Anatolian peninsula, takes the name Dodecanese, which means "twelve islands". There are, in fact, 15 of them, plus a collection of about 150 smaller, uninhabited islets and rocks.

Located at the eastern margin of the Aegean, the Dodecanese Archipelago lies in a transition zone between the Cyclades in the middle of the sea and the Anatolian mainland, where climatic conditions are quite different: hotter and calmer in summer, and colder in winter. The Dodecanese thus benefits from milder winds and a slightly wetter climate than the Cyclades, but not quite as harsh as in nearby Türkiye. Even under the brunt of the driest of summers, when not a single drop of rain falls from the sky between June and September, the Dodecanese *phrygana*—as the Mediterranean scrubland is known in Greece—is dominated by evergreen lentisk, thyme and rock rose, conferring an eye-pleasing green to the islands' landscapes that stand in stark contrast to the yellow stubble typical of the parched Cyclades.

Alas, many Dodecanese islands would be even more ravishingly beautiful had they not been abused by human mismanagement. Vestiges of forests deserving such a name remain only on the islands of Rhodes and Kos, the two largest of the group. Elsewhere, forest coverage, and the benevolent microclimate that trees provide, has yielded to low-lying *phrygana* because of deforestation. Extensive terracing covers island slopes, where grapes and grains used to be cultivated in the past and are now abandoned, left fallow and reinvaded by *phrygana*. Alas, forests are prevented from reconquering the land by multitudes of goats let loose on the terrain, munching on anything alive like a permanent swarm of locusts, with the prodigal blessing of the European authorities.

Nor is the environment below the sea surface faring much better, as we shall see. Nevertheless, the Dodecanese garland of islands still retains a special type of beauty that would justify any effort to restore its native ecosystems to pristine naturalness.

After leaving the dock in Pythagorion, I pointed *Pontoporia* to the south, towards the Arki islets group. The boat sailed fast with the benefit of a Meltemi that was causing a choppy sea surface but was blowing now so much gentler than before. Despite my life spent applying a strict scientific approach to interpreting nature's manifestations, I am unable to avoid attributing intentions to a wind that seems so keen on wishing me drowned and dead one day, and then behaves like my most loving friend the next. This inability makes me fully sympathetic to the ancients' interpretation of the wind's vagaries as the fruit of a posse of quirky gods.

The choppy conditions rendered the encounter with a large pod of common dolphins ever so much more pleasant. The dolphins were jumping clear of the water at their graceful and fast-moving pace and would have been visible even in rougher waters. Perhaps enjoying the steady speed of *Pontoporia* and taking advantage of the bow wave, the dolphins remained in my company for a good 10 minutes, which is a long time for common dolphins. Turning on their sides and glancing up at me from under the bow, displaying the elegant hourglass pattern of their flanks, they seemed to me quite content with their condition, no doubt an improvement from that of pirates in the night of mythological times. I turned on the hydrophone mounted on *Pontoporia*'s hull and was able to hear the chorus of high-frequency whistles, which the dolphins use for communication, mixed with a crackling of biosonar clicks by which they were acoustically exploring their surroundings.

The dolphin sounds were particularly loud on that day, a sign of special excitement that was running through the pod. I have no idea why that might have been. But a more mundane mystery was clouding my understanding of my graceful travel companions. Why are common dolphins so oxymoronically rare? Why were they reduced to inhabit only scattered pockets of the Mediterranean—for example in the Alborán Sea, in waters west of Sardinia, to the west of Greece, off the southern shores of Israel, and here in the waters south of Samos where they are regularly encountered—instead of being more evenly distributed throughout the Mediterranean? Could these few strongholds be places from where they could eventually multiply and spread, or was I witnessing vestigial remains of a formerly broad reign that they have left forever? What is special about the waters south of Samos? One thing is certain: the vast marine expanses that appear homogeneous to our myopic human eyes are, to marine dwellers, a mosaic of radically different conditions, some better to inhabit than others. Understanding the environmental heterogeneity of the ocean, and its sources, is a key to better management and to fostering species recoveries.

As the last of the bow-riding dolphins decided to leave my company, veering broadly towards the deep and away as they always do when the time has come for them to go, I lifted my eyes to contemplate the surroundings. Arki with its retinue of lesser islets, followed by pretty Lipsi, were rapidly streaming by to port. To starboard stood the enticing profile of the holy island of Patmos, dominated by the immaculate cluster of dwellings of the Chora surrounding the imposing monastery of Saint John, gleaming in pure splendour on the island's highest point. Patmos is a site of great importance to me, vying for dominance in my heart with my Venetian birthplace. The temptation of making a stop on the island was almost irresistible. But a few thousand miles remained in *Pontoporia*'s journey before I could allow myself the luxury of returning there.

Just south of Lipsi lies yet another cluster of tiny, uninhabited islands called the Chàlavras. The largest of these, named Makronisi ("big island"), in relative terms as its longest dimension is just over 1000 m, forms a gentle arch open to the south, thereby offering good shelter from the northerly winds that dominate in summer. I dropped anchor there around mid-day.

The islet consists of a large slab made of thin layers of light-coloured limestone, settled on the shallow shelf south of Lipsi like a gigantic book with some of its pages still intact and some partly torn by geological upheaval and perforated into caves and pools, with pieces projecting into the sea to form isolated pillars. The beauty of Makronisi and the clarity of its waters attract scores of tourists aboard yachts large and small during the summer, only to revert to its natural peace when summer ends. Eleonora's falcons nest on top of the cliffs and soar screaming over the turquoise waters in their breathtaking acrobatics, conferring to the place its own special kind of magic.

Turning my eyes from land to sea, I saw that the sea floor appeared strewn with layered rock fallen from the cliffs above, bathed in the most transparent of waters. Just below *Pontoporia*'s hull, the bottom was littered with a conglomerate of broken amphorae petrified by time inside the rock, silent vestiges of a forgotten tragedy. As far as marine life was concerned, I witnessed the same desolate sight I had seen everywhere in the Dodecanese on many previous occasions: almost no algal cover, few or no fish, and rocks covered with a slimy brownish substance. What has happened to life and biodiversity on the Dodecanese seafloor? I remain flabbergasted by the contrast between the beauty of the landscape above the surface and the drabness of the seascape below.

Here some valuable insight was provided by the findings of my colleague Nike Bianchi, a marine ecologist from the University of Genoa, which he published a few years ago in a scientific journal.[12] Nike and his colleagues conducted two ecological surveys slightly more than three decades apart, one

---

[12] Bianchi C.N., Corsini-Foka M., Morri C., Zenetos A. 2014. Thirty years after: dramatic change in the coastal marine ecosystems of Kos Island (Greece), 1981–2013. Mediterranean Marine Science 15(3):482–497. Doi: https://doi.org/10.12681/mms.678.

in 1981 and the second in 2013, to describe the biological communities of the nearshore sea bottom around the island of Kos, two dozen nautical miles south of where I was anchored. They found that about a quarter of the most common species of algae, fish, crabs, and sea cucumbers they had recorded in 1981 were gone in 2013. In their place, new species had appeared, including some that tolerate waters that are either warmer or more polluted, as well as foreign species that had invaded the Mediterranean through the Suez Canal from the Red Sea.

What Nike and his colleagues had documented with this unique replication of surveys was a true phase shift that had occurred in the biological communities of the sea bottom, making their survey sites ecologically unrecognisable in the span of 32 years, a very short one as far as ecological processes are concerned. The broader environmental changes that had occurred during that time had played a role in causing such a shift. Global warming has caused the sea surface temperature of the eastern Mediterranean to rise by almost 2 °C during the past 30 years. There has also been an extraordinary increase in human pressure on the marine environment, including heavy discharge of waste in coastal waters from summer tourism. Furthermore, indiscriminate overfishing has caused imbalances in fish communities and their dependent ecosystems. Finally, the coup de grace was delivered by the invasion of many alien fish species, such as the Red Sea rabbit fishes,[13] rabid grazers of algal covers that leave behind wide spans of bare rocky bottom. All these changes, which are ongoing and likely accelerating, bear the signature of *Homo sapiens* and are unfortunately responsible for much worse consequences than the status of the ecological communities of the Dodecanese seafloor.

It was time to leave the small paradise of the Chàlavras, thoroughly enjoyable despite the poor underwater conditions, and continue *Pontoporia*'s voyage. I hauled up the anchor and proceeded, with the push of a now much tamer Meltemi, to the south along the east coast of Leros. Upon reaching the island of Kalymnos, I decided to anchor in the small, cosy shelter of an eastern cove named Vathy instead of on the south coast in the island's main port.

Kalymniots were once famous for marketing sponges collected from great depths over the eastern and southern Mediterranean. Historically, they harvested sponges through rather dangerous diving practices, at great risk to their health in the best of cases. Today, the sponge trade is no longer economically viable because natural sponges have been replaced by cheaper artificial

---

[13] Rabbit fishes, *Siganus luridus* and *S. rivulatus*, are native to the shallow waters of the Indo-Pacific marine realm, but have become established in the Eastern Mediterranean having invaded it across the Suez Canal, to the extent that they have become a commercially important item for human consumption in many Mediterranean locations. See also https://en.wikipedia.org/wiki/Rabbitfish.

products. Undeterred, Kalymniots engage in many other types of fishing, and their island has the largest and most active small- to medium-size coastal fleet in all of the eastern Mediterranean.[14]

In light of the island's vigorous fishing community, sailing in Kalymnos waters evokes in me the sense of unease I get when discussing the exercise of fishing at sea. I have a huge problem with the subject in its broadest sense, particularly insofar as industrial fishing is concerned. To explain myself, I have to start from the beginning.

To secure the proteins needed for sustenance, humans have transitioned across the millennia from a hunter-gatherer mammal for whom animal-based proteins were derived from wild species to the current condition whereby this component of our diet is largely derived from domesticated animals. It could not be otherwise given the large amounts needed to feed today's human population. Until recently there remained one notable exception to domesticated sources: marine fisheries. It was once believed that only the oceans were big enough to tolerate the massive subtraction of biomass necessary to sustain a human population numbering in the billions. Alas, we now know that this is no longer true. As mentioned before, most fish populations are now severely depleted, and it is urgent for humanity to source its dietary needs from elsewhere, namely, with much greater attention to plant-based solutions.

Overfishing and poorly managed fishing in any form, licit or illicit, are by far the primary enemies of marine ecosystem balance, with a staggering amount of destructiveness from local impacts to a global scale.[15] All other threats to marine ecosystems pale compared with fisheries depletion, even including pollution by plastic that everybody nowadays gets so (rightfully) excited about. However, the challenges to fisheries are not limited to the extraction of living biomass from the oceans, which is bad enough in itself. Many fishing methods cause habitat destruction, and that destruction, coupled with the plastic pollution from abandoned or discarded fishing gear, is responsible for wildlife death long after the fishers have departed. According to a recent estimate, about 2% of the world's fishing gear, including tens of thousands of square kilometres of nets, hundreds of thousands of kilometres

---

[14] Roditi K., Vafidis D. 2019. Net fisheries' métiers in the Eastern Mediterranean: insights from small-scale fishery management on Kalymnos Island. Water 11(7), 1509 https://doi.org/10.3390/w11071509.

[15] Palomares M.L.D., Froese R., Derrick B., Meeuwig J.J., Nöel S.L., Tsui G., Woroniak J., Zeller D., Pauly D. 2020. Fishery biomass trends of exploited fish populations in marine ecoregions, climatic zones and ocean basins. Estuarine, Coastal and Shelf Science 243, 106896 https://doi.org/10.1016/j.ecss.2020.106896.

of longlines, and more than 25 million pots and traps, are abandoned in the ocean annually.[16]

What I find particularly disturbing is that this destructiveness is far from an unavoidable outcome. Good, sustainable fishing is not an oxymoron. In fact, I do not condemn the eating of seafood as a matter of principle only because I am convinced that fishing sustainably is not impossible. However, let us not fool ourselves: the rarity of fishing activities that occur sustainably today is evident. It is too easy for the rogue fisherman to flout the law, despite the potential monitoring afforded by technologies such as the one provided by planet-wide, nongovernmental programs like Global Fishing Watch.[17] I remain astonished that throughout global policy and economic arenas, the industrial fishing sector enjoys a political clout that is vastly disproportionate relative to its economic relevance, a situation that is amenable to facilitating the widespread tolerance of illegal and harmful practices.

Even assuming for a moment, *ad absurdum,* that all fishing practices were legally conducted, I still have a problem with the activity that is of a more ethical nature. The notion that a fisher is entitled to consider a fish that freely swims in the sea as his property makes me very uncomfortable. Given my special relationship with the piscine tribe, I wonder this: if a fisher has the legal right to catch a fish in the sea, and acquire it to eat or sell, why do not I have the same right to will that same fish to be swimming free and undisturbed? I find it difficult to accept that I have fewer rights than another person to decide the fate of a fish, just because that person is a fisher. If we feel comfortable with the notion that a free-ranging fish can be the property of a human, why is it that the act of killing that fish, thereby subtracting it from the commons for personal advantage, takes not only legal but also moral precedence over the act of willing that particular fish to remain a healthy part of the marine environment and therefore part of the global commons?

I am, of course, ready to admit that the act of fishing derives from a natural foraging behaviour developed by early humans who, like other predators, wished to satisfy their need for proteins. Although this condition has changed significantly today, with captured fish no longer an item of subsistence but instead a commodity to be harvested industrially and traded, I am not arguing that fishing activities should disappear overnight from the face of Earth— although such an eventuality would certainly bring about an immense improvement in the health of the oceans. Instead, I would like to offer a

---

[16] Richardson K., Hardesty B.D., Vince J., Wilcox C. 2022. Global estimates of fishing gear lost to the ocean each year. Science Advances 8(41) https://doi.org/10.1126/sciadv.abq0135.

[17] Global Fishing Watch: https://globalfishingwatch.org/.

compromise and accept the practice of fishing—provided that it is conducted sustainably. By sustainably, I mean not only with respect to the maintenance of the target species but also with respect to the role that species plays within its ecosystem, its function as prey, predator, scavenger, or commensal.

I argue that under no circumstances should fish be subtracted from its population at a level of exploitation that would prevent that population from replenishing itself, an outcome that would also adversely impact the survival of the predators who naturally subsist on them. Indeed, I consider a fish that has been caught unsustainably or worse, illegally, as having been stolen—and the fisher who caught that fish nothing less than a thief. Stolen from me and from the commons, also including everyone who, like me, wishes the fish to be swimming free and undisturbed.

All these reflections about the ethical aspects of fishing normally materialise when I come into contact with communities that are particularly versed in the exercise of fishing, as is the case of Kalymnos. Noting that some Kalymniots have established for themselves an unflattering reputation of plunderers during their fishing forays across the Aegean and the wider Mediterranean, I regret that such a reputation may be reflecting negatively on the island's otherwise decent and law-abiding community, hopefully to be soon forgotten in a future climate of this community's improved relationship with Poseidon's realm.

I dropped anchor in front of the lovely little village of Vathy, at the end of a narrow fjord on the east coast of Kalymnos, as a purple full moon was rising rapidly from above the coast of nearby Türkiye. Minutes later, the moon had become bright white and had transformed the parched cliffs around the bay into a breathtaking landscape that seemed made of silver.

Day 14: Kalymnos, Greece

As I ate breakfast in the cockpit before heading to sea again, a flock of elegant gulls I had never seen before started fluttering around *Pontoporia*. They were slightly smaller and lighter than their omnipresent, preposterous yellow-legged cousins. Their magnificent ash-grey plumage, white head and chest, and strikingly red beak told me that a delegation of Audouin's gulls[18]— threatened Mediterranean endemic birds—had come to greet me and to bade me good winds, gracefully flying above and around, and hoping to get some scraps of my breakfast cake in the process.

After pulling out of the fjord, with Kalymnos fading astern, I set out to navigate the easternmost waters of the Aegean through the narrow strait separating Greece from Türkiye, thus Europe from Asia. A tame Meltemi, gusting occasionally to remind me of the wind's penchant for brutality, was pushing *Pontoporia* along the intended route. My goal that day was to cover the many miles to reach Rhodes, the gate to the Levantine Sea, before dark.

The coast of Türkiye, massacred by the unpleasing estate development surrounding Bodrum, streamed by to port, and the Greek islands of Pserimos and then Kos passed to starboard. I continued towards the beautiful, pristine Datça Peninsula, skimming the enchanting double port of ancient Knidos, resolving to return there and explore the site one day.

When the island of Symi came into view to port, I was in Greek waters again, whilst the lights of the city of Rhodes on the northern tip of the island were glimmering in the approaching dusk. One hour later, I entered the Mandraki harbour, sailing between the bronze statues of two fallow deer—a stag to the right and a doe to the left—atop two columns flanking the entrance, right where the Rhodes Colossus once was said to be standing, legs astride the harbour opening. Fallow deer still inhabit the island's forests and are likely the reason for the island's ancient name of *Elafousa* (Greek for "a place of deer").[19] I decided that the welcome provided by the graceful bronze ungulates was rather more becoming and elegant than what I imagined must have been like, in the old days, to enter the harbour by sailing between the Colossus' legs and under its dangling pudenda.

---

[18] The Audouin's gull, *Larus audouinii*, is currently assessed as "Vulnerable": https://tinyurl.com/546562bb.

[19] From the Greek word Ἔλαφος, deer. Sadly, the Rhodes' population of fallow deer was clobbered by devastating fires in July 2023 that ravaged about 135,000 hectares of forest on the island. See, for instance: https://tinyurl.com/33ucmn6s.

6

Levantine Sea: From Rhodes to Paleochora

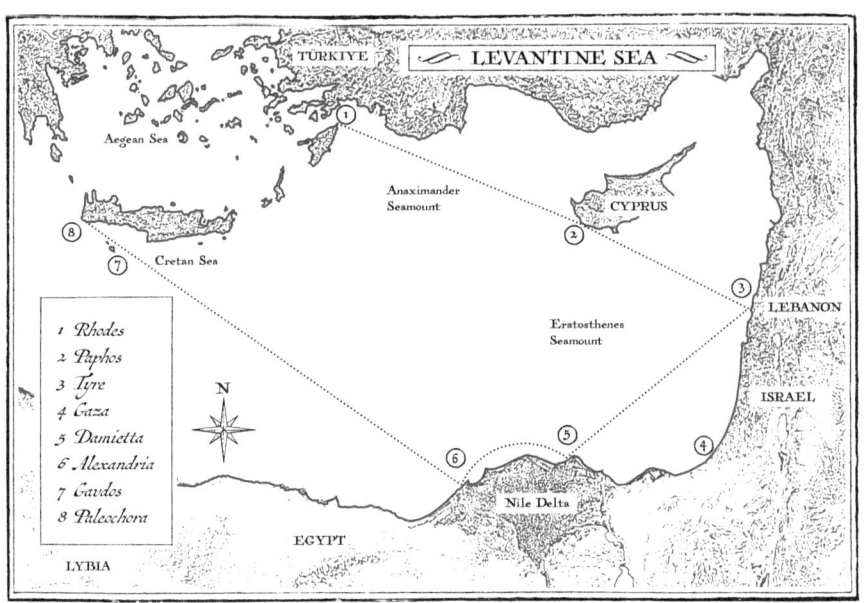

Illustration by Massimo Demma

G. Notarbartolo di Sciara, *Sailing Across a Wounded Sea*,
https://doi.org/10.1007/978-3-031-54597-9_6

Day 15: Rhodes, Greece

Pontoporia and her crew needed a whole day of rest in Rhodes, my last stop in Greece, before forging into Middle Eastern waters. It was an excellent opportunity for man and boat alike to catch some breath after sailing non-stop for 9 days and having covered about 660 miles since I left Otranto at the southeastern tip of Italy. *Pontoporia* needed restocking with food, water, and fuel, particularly in the expectation that my meanderings across a mostly windless summertime Levantine Sea would require extensive use of the auxiliary engine.

History has deposited layers of successive civilisations on top of each other here on the main island of the Dodecanese, making it possible to travel across millennia with the wings of the mind whilst remaining in the same place. Boarding with my imagination the stunning Dorian ship carved in bas-relief in the Greek classical period within the rock face overhanging the entrance to the Acropolis in Lindos (a seaside town about 50 km south of the main city), I would sail across successive Hellenistic and Byzantine rule around Rhodes' medieval walls which the Knights Hospitaller fortified the town with, and then slalom around the minarets of Rhodes' many mosques testifying to the long Ottoman domination. But I got the strange sensation of walking across even more familiar territory when strolling around the city's harbour, because the main buildings lined up along the boulevard looked unambiguously Italian. The Dodecanese had been under Italian rule a century ago, from 1912, when the archipelago was conquered from the Ottomans, until 1947. During that period, Italy left an architectural legacy that still stands today, with a collection of buildings subsequently adapted by independent Greece to serve the purposes of its administrative machinery. Interestingly, one can trace Italy's involution during that period of her history from the post-Great War years into the nation's toxic embrace of fascism by noting the transition from the eclectic, vibrant architectural fashion of the earlier Dodecanese buildings—where Venetian, Islamist and Gothic elements were combined in an intriguing mix of styles—to the stern, almost brutal unfriendliness of the later rationalist edifices.

A brief stroll from *Pontoporia*'s berth along Mandraki's outer pier amongst throngs of holidaymakers brought me first in front of the white, imposing entrance to the "Néa Agorà" ("Nuovo Mercato", the market hall), designed by Florestano Di Fausto and completed in 1926. Inside the tree-covered internal court, making my way across a bustling crowd of tourists mixed with locals populating the many coffee bars, tavernas, jewellery, and gift shops, I could reach the food stalls where I would get all the provisions needed for the coming weeks at sea. Having exited the market and progressing northward by a few hundred metres along a spacious, tree-lined avenue, I passed along the building of the "Limenarkion" (Coast Guard), then the Palace of the Government, the Theatre, the Lido, finally reaching at the very northern tip of the island the Aquarium, a small architectural jewel completed by Armando Bernabiti in 1935. Admittedly, I found the displays of the aquarium and annexed museum to be a bit lacklustre—this being the fate of most aquaria anyway, for it is impossible to do justice to the magnificence of the ocean with human-built replicas, not even if making them cost millions—but I could not avoid appreciating the intention of showcasing a sample of the representatives of the local marine fauna for the benefit of the public. Here, minor characters, like reef fishes and octopuses, were kept alive in the small tanks. In contrast, stuffed specimens of the larger species, including a few sharks, a spinetail devil ray and even a Cuvier's beaked whale, were conserved in the annexed museum, which looked more like an old Wunderkammer intriguingly displaying its guests frozen in various postures of deathly contortion.

But *Pontoporia* needed some attention before the next bout of navigation, so I walked back from the Aquarium and spent the rest of the day fixing things and preparing for the next morning's departure. As the sea breeze was dying out with the descending night over the Mandraki, and the town's evening lights were coming alive, I found it enjoyable to get busy inside the boat's familiar microcosm and let sleep conquer me in the harbour's motionless, albeit somewhat buzzing embrace.

Day 16: Rhodes, Greece

Having departed from Rhodes in the morning after having bid goodbye to the stately bronze deer couple guarding the Mandraki, *Pontoporia* was proceeding at a good pace in a south-easterly direction, propelled by the auxiliary engine in the total absence of wind and the flattest of seas. By noon, the heat of the day was becoming a vexing companion. Despite the proximity of the Aegean Sea, the surroundings here could not have been more different, with the influence of the Meltemi having become irrelevant and a water depth that a mere dozen miles from the coast was already greater than 2000 m and increasing as I proceeded eastward.

Welcome to the Levantine Sea, where the Mediterranean, typically an elongated marine buffer separating Europe from Africa, abuts the westernmost side of the Asian continent. I was soon going to transit away from the Western world where I had been until hours ago. The notion of being about to cross a cultural boundary was also reinforced by that of sailing in a seascape that was so different from that of the Aegean. I had the sensation of floating in waters irradiating a sense of tropicality, above a deep abyssal plain disseminated with hidden, yet towering seamounts. Here in the Levantine Sea, marine species that have become rare in the west, like monk seals, or have even been extirpated from there, like guitarfishes and many species of rays, are still present. I am not sure why that is because I do not think that fishing pressure is lesser here than elsewhere in the Mediterranean. That said, today native species must contend their space to a legion of scaly invaders from the Indian Ocean marine realm, barrelling into waters made warmer by the man-made disruption of Earth's climate through the gaping doorway of the Suez Canal, cut by human hands across the deserts of Egypt. This means that the fate of marine biodiversity here, as we had known it in the past, is as uncertain as it can be.

Pontoporia was soon pushing forward across the flat surface over waters deeper than 3700 m—the deepest ever I had been sailing over since leaving Venice. From now on, my navigation was going to take a different pace, a novel rhythm being required from both man and boat, more demanding of our respective navigational endurance, with fewer opportunities to stop in harbours, relax and visit the lands, and more time spent at sea on longer passages. On the starboard side laid the vast expanse of the Levantine Sea; to port stood the mountainous coast of Anatolia, past the bay of Fethiye, and then the town of Kaş right across from the island of Kastellorizo, the last of the Dodecanese and Greece's easternmost outpost, about 15 miles north of my course.

Sitting in the cockpit under the awning, seeking protection from an unforgiving sun with no hope of getting much cooling from the warm sea below the keel, the monotony of the progress and the droning noise of the motor had been driving me towards a state of drowsiness. There was not much to pay attention to anyway, for the ship traffic was low in that area, with no vessel appearing over the horizon for hours. It was in such a somnolent state that my eyes fell, almost by chance, on the water a few metres across from the gunwale, where two large, ghostly figures appeared under the clear surface, perfectly matching *Pontoporia*'s speed as they were gliding forward with no apparent effort. For an instant, I imagined that such an unreal presence could have been appearing to me only in a dream. But soon adrenaline kicked into my veins, and I realised that I was being blessed by the visit of a mother Cuvier's beaked whale with her calf.

The big mama was at least 5 m long, her light-coloured forehead shading off into a brownish body; her skin marked all around with scratches and pockmarks obtained in god-knows-what wild adventures; her toothless, upturned

Fig. 6.1 An adult female Cuvier's beaked whale unobtrusively surfaces showing her very light dorsal colouration. Her dorsal fin, situated at around two-thirds of her body, is still underwater (Illustration by Massimo Demma)

mouthline reminding me of a bizarre smile, one that has warranted the species the alternate common name of "goose-beaked whale". The calf—about 3 m long, its body uniform grey and unblemished—was swimming at almost touching distance from its mother as if glued to her flank and cautiously maintained itself on the opposite side of the boat. As the ever-diligent naturalist, it took me no more than 2 seconds to avert my sight to take note of the encounter's position from the cockpit GPS, so that I could provide complete data on the sighting to colleagues currently researching the cetacean fauna of this area. Still, when I lifted the eyes again to have a better look at the strange couple, they had vanished. Probably gone again to the depths these beings feel so comfortable in (Fig. 6.1).

These mysterious cetaceans, a whole family of them—called Ziphiids, or beaked whales—include 24 of the weirdest marine mammals one can think of, widely distributed across the world's oceans but regularly represented in the Mediterranean only by a single species, the Cuvier's beaked whale. Some of these whales are so rare and so little-known, that they have been named and scientifically described only on the basis of a few bony remains found stranded on the world's remotest beaches. Ziphiids share the habit of living in very small groups, although most details of their social lives largely remain unknown. One aspect that makes them special is their ability to swim down to extraordinary depths to hunt squids that can be found only there. Cuvier's beaked whales, in particular, hold the mammalian dive record, being able to reach, and likely exceed, depths of 3000 m and for a duration above 2 h. These animals are so comfortable at depth that it has been calculated that they spend, on average, more than 85% of their life away from the surface: a remarkable adaptation to the watery medium considering that, like all mammals, they need air to breathe.[1] This behaviour, however, brings with it the added benefit, not negligible in modern days in the Mediterranean, of keeping them away from the ships' keels and propellers in what has become one of the world's most trafficked seas.

It was not by chance that I had encountered our first Cuvier's beaked whales here, to the east of Rhodes, given that these waters—amongst the deepest in the entire Mediterranean—are included in the "Hellenic Trench IMMA", which was identified for its global relevance to Cuvier's beaked whales and sperm whales.[2] I might also have run into them when sailing across the waters of the eastern Ionian Sea, another beaked whale hotspot in the Mediterranean region, before entering the shallow waters of the Inner Ionian Sea Archipelago. Due to their low natural density, it probably didn't happen there because these cetaceans are rare, even in the known hotspots of their occurrence. This, combined with their shyness and habit of spending most of their life below the surface, makes the chance of encountering them very low.

One would think that marine mammals like beaked whales, their range limited to the open sea, naturally wary of human presence, and subsisting on squid, a food source so deep that it is beyond the reach of most other mammalian predators, would be able to enjoy a relatively carefree existence even in a human-infested region such as the Mediterranean: not so. In addition to the "usual" gauntlets that cetaceans must experience, like drowning in illegal driftnets, being hit by the bow of a ship, and having their gullet obstructed by plastic trash that should not have been in the water in the first place, rather surprisingly beaked whales were found to be quite vulnerable to the powerful sonars of military vessels, with tragic consequences. It was in the 1960s that puzzling reports started to appear of groups of beaked whales found stranded all together, dead or dying, without any visible sign that could help to make sense of their demise. The onset of these happenings coincided with the novel use, in those years, of powerful antisubmarine sonar by several of the world's navies. One of the first of such incidents, if not the first, occurred along the coast of western Liguria in the north-west Mediterranean in November 1963, when about 15 Cuvier's beaked whales were found stranded with no external signs of injury over a stretch of coast tens of kilometres long, in concomitance with the presence in those waters of several vessels from the US and UK navies. At the time, no connection could be made between the carnage and

[1] Shearer J.M., Quick N.J., Cioffi W.R., Baird R.W., Webster D.L., Foley H.J., Swaim Z.T., Waples D.M., Bell J.T., Read A.J. 2019. Diving behaviour of Cuvier's beaked whales (*Ziphius cavirostris*) off Cape Hatteras, North Carolina. Royal Society Open Science 6: 181728. https://doi.org/10.1098/rsos.181728.
[2] Hellenic Trench IMMA: https://tinyurl.com/3vm9tyu9.

the naval manoeuvres.[3] Definite proof of such a relationship had to wait till 1996, when a powerful new sonar, tested by a NATO research vessel in the Ionian Sea off the west of Peloponnese, in Greece, caused the stranding of 13 Cuvier's beaked whales.[4] Since then, several other incidents have been reported from many parts of the world, involving collectively the deaths of hundreds of beaked whales belonging to several different species and establishing beyond any reasonable doubt a cause-effect link between the naval activities and the whale deaths. Today, very few scientists believe that the root cause of the damage to the whales can be direct harm from being hit by acoustic energy, powerful as that might be. More plausibly, the whales, for some still poorly understood reason, seem to panic when they hear the sonar pulses during their deep dives, and rush to the surface without decompressing properly. This, in turn, provokes damages to their internal organs with associated haemorrhages, bringing so many of them to their demise. Unfortunately, the Mediterranean Sea, in addition to being important economically, is also a major strategic theatre for the world's navies, just as it was during the times of the wars between Greece and Persia; with the main difference, as far as the beaked whales are concerned, that Xerxes' fleet did not use powerful sonars in the exercise of warfare. This condition has conservation consequences, and the status of Cuvier's beaked whales in the Mediterranean has been assessed as "Vulnerable" in IUCN's Red List[5] mainly because of the potentially devastating effects of noise on their survival.

Thinking with a sense of guilt about the plight human behaviour is causing these shy giants, I felt gratitude for the emotion received from the appearance so near *Pontoporia* by this mother and her baby. I wondered whether the encounter had been casual or intentional. I suspect that Mama *Ziphius* had decided to surface near me and check me out deliberately. Many years ago, a

---

[3] Littardi V., Rosso M., Wurtz M. 2004. Enquêtes historiques (1900–1966) sur les échouages de *Ziphius cavirostris* G. Cuvier, en Mer Ligure. Rapports de la Commission Internationale pour la Mer Méditerranée. 37:388.

[4] Frantzis A. 1998. Does acoustic testing strand whales? Nature (London) 392:29.

[5] The Cuvier's beaked whale, *Ziphius cavirostris*, is assessed as Vulnerable in the Mediterranean Sea: https://tinyurl.com/2yvn9whs.

similar episode had occurred to me off northeastern Sardinia, also involving a cow and calf pair. Cuvier's beaked whales are known to appear near boats by very unobtrusively surfacing from behind the stern. They will never stop being, for me, a source of wonder.

After the whales' disappearance, my progress continued eventless through-out daylight over the calm Levantine waters, seamlessly merging into a peace-ful night. Whilst the windless daytime was domineered by the blinding light and heat of the mighty sun, the arrival of the night, heralded by the appear-ance of Jupiter above the horizon, brought the relief of coolness. But the real treat was not just of a thermal nature. Despite the still intense light of a moon heading towards its last quarter phase, the canopy of an incredibly starry sky I was navigating under, as *Pontoporia* was cutting the black waters, was a tan-gible reminder that my motion was not only over the surface of our planet but most notably across a greater cosmic dimension. A deep aperçu inducive of existential wondering, increasingly denied by humanity's ubiquitous obses-sion with flooding its night landscapes with artificial lighting.

Day 17: Levantine Sea

Slightly after midnight, I was sailing on top of the Anaximander Seamounts, underwater volcanoes aligned along a contorted ridge rising from a deep-sea bottom around 3000 m deep to a summit 680 m below the surface. The calm sea, glittering under the soft starlight, was giving no obvious sign of the tor-mented roughness of the seafloor underneath, which provided habitat and shelter to who knows what mysterious abyssal creatures. *Pontoporia*, my pri-vate little spaceship, was proceeding in the darkness safely carrying me and my microcosm inside her shell, a minute parcel of human habitat made of wood, metal, and canvas, granting my frail being a safe passage in a world where I would not survive long if left to my own biological devices.

As the sun rose again after a quiet night, changing the scene to a new windless, fierily hot day, the Cyprus landmass eventually became distinctly visible at the horizon. The visibility was good, so it took me several more hours to approach my destination. By then, however, the proximity of Cyprus had generated a pleasant breeze blowing towards land, a welcome change allowing the engine some rest and replacing the hum of the motor with the refreshingly flushing sound of the water on the keel and the wind on the sails.

In front of us, encompassing a large portion of Cyprus' west coast, laid the "Akamas and Chrysochou Bay IMMA", established there due to the presence of caves still used by a small number of monk seals to rest and to breed.[6] Cyprus joins Greece and Türkiye as one of the only three Mediterranean nations where monk seals are still known to reproduce, and this is apparently also happening elsewhere in Cyprus: in the "Akrotiri Peninsula IMMA", a very small area in the south,[7] and in the "Northern Coast of Cyprus IMMA" in the north, currently under Turkish control.[8] However, the species is here represented by minimal numbers, and its status is more uncertain than in the other two larger countries.

The Akamas and Chrysochou Bay are also listed as an "Ecologically and Biologically Significant Area" by the UN Convention on Biological Diversity, not only because of the monk seal habitat it contains but also because of the importance of the beaches for sea turtle nesting. Furthermore, the area also includes one of the finest examples of Vermetid reefs, peculiar coastal formations made not of rock but of concretions of snails cemented together by calcareous red algae. This property allows such reefs to be named "bioconstructions".

Just to the south of that IMMA, an indentation of the coast harbours the town of Paphos with her impressive ancient ruins visible along the shore from the harbour, where I landed mid-afternoon and docked in the small marina.

Day 18: Paphos, Cyprus

I was planning to cover a lot of water across the Levantine Sea, so there was not much time allowed to explore Cyprus, despite the recent knowledge that monk seals have been seen on the island with increasing frequency. Evidence that the small Cyprian colony of the pinniped, believed to be less than 20 strong, still produces pups bodes well for the future of monk seals in Cyprus. So, after a restful sleep inside Paphos' breakwater and the customary visit to a nearby bakery where I was able to stock *Pontoporia*'s galley with koulouri bread covered with sesame seeds, as well as very interesting cheese-filled pastries named *flaounes*, early in the morning I cast off from the pier and took to the sea again, heading for Lebanon. This would be another cruising day promising to bring more calm weather and heat.

In the early afternoon, as the silhouette of Cyprus was starting to fade to port in the midday haze, I had the Eratosthenes Seamount alongside, about

---

[6] Akamas and Chrysochou Bay IMMA: https://tinyurl.com/bdfvmz3b.
[7] Akrotiri Peninsula IMMA: https://tinyurl.com/496drf8z.
[8] Northern Coast of Cyprus IMMA: https://tinyurl.com/2s3aht98.

30 miles to starboard, its summit 690 m below the sea surface. The underwater mountain is a towering, isolated massif rising from the surrounding abyssal plains deeper than 2000 m, and quite a morphological feature of the region. However, it is one that humans tend to ignore just because it is hidden underwater. Alone in the middle of the Levantine Sea, this king of all Mediterranean seamounts looks like a mesa mountain with a vast, 4500 km^2 wide plateau which received its flattened shape during the Messinian Salinity Crisis, about six million years ago, when its top remained above sea level for a few hundred thousand years, and was consequently eroded by weather agency. Oceanographic expeditions made in 1998 and later in 2012 with submersible probes provided rare glances at the intriguing communities of living organisms that have established their hold on the top of the Eratosthenes Seamount, including deep-sea corals, sponges, molluscs, and echinoderms. This condition had transformed the seamount into a submarine oasis of life surrounded by an abyssal biological desert. The biota found on the top of the seamount is further enriched by cracks and vents through which heated water from the inner crust outflows into the surroundings. These passages create the conditions for the flourishing of unique biological communities made of small clams, tubeworms, sea urchins, and crabs.[9] Intriguingly, the findings by the oceanographers were not limited to natural discoveries; videos collected through remotely operated vehicles also documented the presence on the plateau of archaeological remains such as amphorae and other human objects from Greece, Rome, and the Near East, silent witnesses from human tragedies consumed across millennia in these waters hundreds of metres above.[10]

In 2006, the General Fisheries Commission for the Mediterranean declared the Eratosthenes Seamount a FRA to protect its delicate bottom habitats from the potential ravages of deep-sea trawling. Alas, economic interests even greater than fishing got in the way. The Levantine seafloor was recently found to contain significant reserves of natural gas. Cyprus, not to miss out on the bonanza, promptly extended its jurisdiction over a large area by declaring its Exclusive Economic Zone, including the Eratosthenes Seamount.

At a meeting organised in Malaga in April 2014 by the Secretariat of the Convention on Biological Diversity to identify "Ecologically or Biologically Significant Marine Areas" (EBSA) in the Mediterranean, representatives from the Republic of Cyprus fiercely opposed even the slightest mention that the

---

[9] Mitchell G., Mayer L., Bell K.L.C., Ballard R.D., Raineault N.A., Roman C., Ballard W.B.A., Conwell K., Hine A., Shinn E., Dimitriadis I., Bogdan O. 2013. Archaeological discoveries of the Eratosthenes Seamount. Oceanography 26(1):36–41.

[10] Cornwell K., Opaiț A., Ballard W.B.A. 2013. Exploration of Eratosthenes Seamount - a continental fragment being forced down an oceanic trench. Oceanography 26(1):42–43.

Eratosthenes Seamount might be considered for such listing by virtue of the intriguing biological communities that had been described to live on its top. Soon after, the whole area was parcelled out into 13 seafloor blocks for natural gas exploitation, including and surrounding the venerable Seamount, named after the equally venerable Greek astronomer, with gas extraction implemented in a field ironically named "Aphrodite" by the concerned oil companies: a tribute to the Greek goddess of sensual love and beauty whom the legend says was born from the Cyprian Sea's foam, and a testimony to the dubious sense of humour of the hydrocarbon industrialists of the island.

I was now sailing at an easy gait towards the coast of Lebanon in the mid-afternoon, as a pleasant sea breeze generated by the heating of the not-so-distant land had filled *Pontoporia*'s sails and brought respite from the heat of the day.

It was then that a school of dolphins came again to visit me. Like ghosts, the playful mammals had appeared from nowhere and were now happily riding the wave created by *Pontoporia*'s advancing bow. It was immediately apparent, however, that these dolphins were something different from what I had encountered so far. Their drab, faintly marked and mottled grey colouration was not too different from that of bottlenose dolphins. Still, they were shorter and slenderer, with a relatively taller and more triangular dorsal fin. Most of all, they had an entirely different "face": gone was the bulging, rounded front, the so-called "melon" sported by the more conventional dolphin species, replaced by a slanted, receding forehead conferring to these dolphins a sort of archaic, almost reptilian mien. Some also had distinctive white lips, which showed well when I had positioned myself on the bow—having left *Pontoporia* steering herself with the help of the autopilot—to enjoy the spectacle of the bow-riding creatures nimbly jostling for the best position in front of the boat (Fig. 6.2).

When describing the differences between the bottlenose, common, and striped dolphins I had encountered earlier in this voyage, I had already noted how different the dolphin species appear once one learns how to distinguish them. If you pay attention to the right clues—the colour of the body, shape, and position of the dorsal fin, shape of head, behaviour, and so forth—differences amongst species soon appear monumental, not unlike when you learn

Fig. 6.2 Two rough-toothed dolphins from a group of 11 surface in the calm waters of the Eastern Mediterranean (Illustration by Massimo Demma)

to distinguish all the different songbird species that visit your feeder or all the species of gazelles and antelopes that you can encounter when visiting the African savannah. Identifying species is not a pedantic exercise for taxonomy-obsessed geeks: different species have different needs, different behaviours, and different vulnerabilities, and the understanding of such differences is vital to effective conservation—in addition to being interesting traits to be observed, by themselves.

All the characters observed in that day's encounter allowed me to identify my visitors as "rough-toothed dolphins"[11] (a rather ill-chosen common name, in addition to being a bit of a tongue-twister, pointing to the almost invisible and likely irrelevant coarseness of their teeth's enamel). These intriguing mammals are considered brilliant characters even within the Delphinine tribe, and are often attracted by boats and humans. They are also reportedly amenable to interacting with swimmers at sea, a situation that most other dolphin species usually abhor. Some older evidence of these dolphins' behavioural peculiarity comes from studies made in captivity. A female rough-toothed dolphin kept in Hawaii's Sea Life Park named Malia, who was rewarded with tasty fish morsels only when performing novel behaviours, was able to reap one reward after another by quickly understanding what was required from her, coming up with increasingly imaginative performances.[12] Circumtropical in their world distribution, rough-toothed dolphins in the Mediterranean are rare but have been recently observed with some regularity in the eastern portion of the basin, where they are now considered to be represented by a small, threatened population. I remember making one of my life's most memorable cetacean sightings when, just east of the Strait of Sicily back in 1985, during an expedition organised by an old friend, the late William (Bill) Watkins from the Woods Hole Oceanographic Institution, our vessel was surrounded by a huge herd of about 160 rough-toothed dolphins.[13] We could not believe our eyes at the sight of the strange animals darting right and left around our boat because, at the time, the record of these dolphins from the Mediterranean was only based on very sporadic occurrences, and the species was not considered regular in the region. The vessel we were on was equipped with state-of-the-art acoustic recording instrumentation, and we could document features of rough-toothed dolphins' whistles that turned out to be species specific.

Strange things involving rough-toothed dolphins have happened in the Levantine Sea in recent decades. In March 2006, a herd of about 40 came into the port of Haifa in Israel and remained inside for a whilst, apparently in good health but with no obvious motivation for such odd behaviour. It is unusual for groups of cetaceans to come deliberately into trafficked, polluted, and noisy harbours unless they have a good reason for doing so, which in the case of the Haifa incident was unfathomable. More problematic was an episode on 10 March 2010 in Cyprus, when 21 rough-toothed dolphins were stranded

---

[11] The rough-toothed dolphin, *Steno bredanensis,* is assessed in the Mediterranean Sea as "Near Threatened": https://tinyurl.com/4sxfd673.

[12] Pryor K. 1969. The porpoise caper. Psychology Today 3(7):47–49.

[13] Watkins W.A., Tyack P., Moore K.E., Notarbartolo di Sciara G. 1987. *Steno bredanensis* in the Mediterranean Sea. Marine Mammal Science 3(1):78–82.

alive on Laidy's Mile Beach, just south of the port of Limassol on the east side of the Akrotiri Peninsula. Compassionate bystanders pushed the animals back to sea, and no dolphin casualties were reported. Live strandings of cetaceans are very rare in the Mediterranean, and dolphins normally do not throw themselves on beaches just for fun. In the Laidy's Mile Beach episode, the animals appeared to be disorientated, possibly scared by some unknown agent, such as from the intense cycles of seismic surveys conducted in those years by the oil and gas exploration industry. The high-energy noise irradiated into the sea by airguns, specialised contraptions towed by the industry's research vessels during seismic surveys, is known to cause cetaceans to leave the area in the best of cases; episodes of disorientation and strandings are also known, and this would seem the most straightforward explanation for the Laidy's Mile Beach episode, despite local "experts" assurances that there was no temporal coincidence between the industrial activities at sea and the dolphin stranding event.

I was delighted at seeing these dolphins frolicking under *Pontoporia*'s bow. The dolphin habit of rushing in front of boats to surf in the wave created as the vessel advances have endeared these animals to sailors for time immemorial: a rare opportunity for humans to interact with wild mammals that does not happen with wild terrestrial mammals in analogous situations. Unfortunately, we know of instances in which the animals' confidence is betrayed by hoodlums who take advantage of the dolphins' confidence by harpooning or damaging them otherwise. For the better human types, contact can occasionally be made when dolphins ride the bow wave. As noted earlier in these pages, they often place themselves in a group in front of the advancing boat, and they are so busy jostling for the prime positions that they do not give the impression of caring much for the humans who may be watching them from above. Sometimes, though, mainly when their number is small, their attitude is different, and they give the impression of being aware of the human presence and even interested in it. One of these rough-toothed dolphins in particular, likely an older individual with mottled black and white lips, was placing itself again and again in a position right below me, turning on its side so that it could look up and we made eye contact—making that spark of intelligence absolutely evident. I had so many questions I would have liked to ask, but all I could do was turn on the amplifier and listen to the dolphins' whistles. Sure enough, they had the very typical rough-toothed dolphin stepping-up character.

The dolphins stayed with me for a few more minutes and then dissolved in a second into the dark blue water as they often do, leaving me to proceed alone over the smooth surface as the sky was getting darker. No other creatures happened to cheer up my progress in the warm night that soon followed.

Days 19–20: Levantine Sea

At the break of dawn, the Asian landmass had become visible in front of me as the sun rose from behind Mount Hermon's heights. Unsightly high-rise buildings had appeared just inland from the ancient city of Tyre, where I was bound and where I docked early that morning, inside the old town's small fishing harbour.

Tyre had been a Phoenician seaport of significant importance since 2000 BCE and for no less than a few millennia. The harbour where I docked used to be on an island, albeit very close to the mainland coast. In 332 BCE, Alexander the Great laid siege to the city, and to facilitate its conquest he built a causeway, which changed the island into a peninsula as it is today. Tyre's economic importance started declining after that conquest, and it is said that some of the Tyrians ran from their homeland to Carthage, in what is today's Tunisia. Today a UNESCO World Heritage site, Tyre is yet another of the many places I have visited in this journey where the layers superimposed on top of each other by time—from Roman to Hellenistic to Byzantine to Islamic—have become so commonplace in this part of the world that one has to be careful not to take it all for granted, and stop from time to time to think and remember how fleeting can be the societal cocoon protecting us at the time in history we live in. Tyre is the perfect case in this instability having seen its regional importance ebbing and flowing many times depending on who called the shots in the region at any particular time. As a case in point, Tyre has sadly ended today in a condition of near irrelevance compared to the past glories, made worse by the conditions of a country in deep economic and social troubles. Even the marbles in Tyre's archaeological site, the last memento of ancient opulence, are at risk of decay due to airborne pollution from the surrounding unchecked urban sprawl.

I had visited Lebanon 10 years before, invited to consult with the Italian Embassy to support the development of marine conservation in the country. At that time, I had examined the conditions of one of the two MPAs in Lebanon, the "Tyre Coast Nature Reserve": a nationally declared marine and land protected area within walking distance from where I was, to the south of the city. The area includes a large sandy beach, known as an important nesting site for loggerhead and green turtles. The protection it affords to the natural habitats is minimal, with slightly more than a 100 km^2 of marine water included within its boundaries, and a mere 4 km^2 of land, surrounded by heavily urbanised areas including a refugee camp. Not much indeed, but one has to grab any existing opportunity to decrease human pressure in overpopulated situations such as in southern Lebanon, and defend the area with all possible means. Accordingly, and despite its small size, the "Tyre Coast Nature Reserve" was declared a "Special Protected Area of Mediterranean Importance" within the framework of the Barcelona Convention.

I still had enough fuel to continue my sailing to Egypt so I could dedicate my time to stockpiling *Pontoporia*'s storeroom with Lebanese food. Lebanon's luck might be rollercoasting between ill and good depending on the vagaries of history, but the one thing it has eternal is its food. As is already my habit, I wandered across old Tyre's narrow streets guided by my nose, zeroing in like a predator onto a trove of delights destined to spice up—or sweeten depending on circumstances—many a solitary hour at the helm of *Pontoporia*. So, in came aboard bags full of bread types such as pita, saj, fatayer, serpentine mouwarkas, to be tasted with that superb blend of herbs, sesame, and sumac called za'atar, mixed with the extra virgin oil I still had plenty of from Otranto. And it also came with jazarieh (shredded pumpkin cooked in sugar syrup), nammoura (dough semolina) and—*dulcis in fundo*—glorious kunāfah, layers of thin kataifi filled with cheese cream, flavoured with cardamom and orange zest and drenched in a syrup blended with lemon juice and bitter orange blossom water. And may Hestia, the ancient feasting goddess, forever protect me from hyperglycaemia.

Just past noon, I cast off from Tyre and took to the sea again, heading southwest towards the Nile Delta area. I was now sailing in the far south-eastern corner of the Mediterranean Sea, bound by the Israeli and Palestinian coasts to the east and Egypt to the south. The afternoon sea breeze filled *Pontoporia's* sails and helped her to eat mile after mile until well after sunset. As I proceeded obliquely, my distance from the Asian landmass progressively increased as it decreased from the African continent. Meanwhile, with darkness descending from the east, the glow from the busy Israeli coast was becoming apparent just beyond the horizon, enhanced by the bright lights of the Tamar and Leviathan gas terminals. It was hard to shake off the unpleasant sensation that no wilderness was left when sailing across these waters. And yet, one should never stop at first impressions. Despite being busy with human activities, Israeli coastal waters are rich in marine fauna, as reported by colleague and friend Aviad Scheinin, head of the Israel Marine Mammal Research & Assistance Centre. For decades, Aviad and his colleagues have kept a vigilant eye on the presence of marine mammals and other exponents of marine megafauna off the Israeli coast, largely consisting of bottlenose dolphins but also including the occasional monk seal and a community of common dolphins resident off the southern part of the coast. In 2016, thanks to a large extent to Aviad's work, an Important Marine Mammal Area was identified there, with the name "Coastal Shelf Water of the South East Levantine Sea IMMA", to highlight the presence of this common dolphin tribe, now rare in the Mediterranean.[14] During his diligent patrolling, Aviad did not miss the occurrence off the Israeli coast, on 8 May 2010, of a visitor of exception to the Mediterranean east shores: a grey whale, approximately 13 m long, arrived there after an extraordinary journey from the North Pacific, which is the only region that today is inhabited by this cetacean species.[15] The same wanderer was sighted again off Barcelona by Manel Gazo, another colleague, about a month later—before disappearing without leaving any further trace.

Early in the morning the following day, I was as far from land as I would get during this leg. Well beyond the horizon to the portside, about 70 miles

---

[14] Coastal Shelf Water of the South East Levantine Sea IMMA: https://tinyurl.com/5b6fyx8u.

[15] Scheinin A.P., Kerem D., MacLeod C.D., Gazo M., Chicote C.A., Castellote M. 2011. Gray whale (*Eschrichtius robustus*) in the Mediterranean Sea: anomalous event or early sign of climate-driven distribution change? Marine Biodiversity Records Vol. 4; e28. https://doi.org/10.1017/S1755267211000042.

Fig. 6.3 A spinetail devil ray is swimming fast at the water's surface. Looking at the whitish scars near the tip of the right pectoral fin, this ray is likely a recently mated female. Males leave such scars on the delicate skin of the female when they bite the female's fin to strengthen their grip during mating (Illustration by Massimo Demma)

east of my course, was the Gaza Strip, the narrow Palestinian enclave carved along a 41-km-long tract of Mediterranean coast between Israel and Egypt. Looking in that direction, my thoughts went to Mohammed Abudaya, another colleague and friend, and to all the hardships he and his family must endure, caged in that cramped land under the scourge of madmen in charge on both sides of the border.

Mohammed teaches ichthyology at Gaza's National Research Centre and is an expert in studying and conserving sharks and rays. I contacted him in February 2014, when the news travelled worldwide that dozens of large rays had been found stranded on the beach of Gaza City. A glance at the reports told me quite a different story from what had appeared in the media. If cetaceans are unfortunately known to mass strand at times, this phenomenon is unheard of regarding rays. These cartilaginous fishes relatives to sharks lack the cohesive social behaviour that brings some cetacean species, such as pilot whales, false killer whales, melon-headed whales, and sperm whales, to keep tight into their extended family group when inadvertently running into shallow waters, eventually into a deadly trap where they are unable to free themselves from. Although a few species of rays exist that have the habit of living in large groups, evidently their behaviour, including the ability to swim in shallow water, must protect them from the ruinous consequences that are so lethal to the cetaceans concerned (Fig. 6.3).

The rays in the Gaza incident were spinetail devil rays,[16] a species very well known to me because I have dedicated a substantial portion of my professional life to the study and conservation of these fishes, which also were the subject of my doctoral dissertation in what seems to me the night of times. Spinetail devil rays are 1 of 11 species in the family Mobulidae. These are quite large compared to most rays, but two—commonly known as mantas or manta rays—are truly gigantic. Mobulid rays, or devil rays which is the same thing, are unusual in many respects. Besides being extra-large, they have adapted to feed on zooplankton and small schooling fishes, which they catch in their large mouths by swimming in the water column and often at the surface. This feeding habit has transformed them from bottom-dwelling fishes, like the more "conventional" rays they have derived from, into highly efficient pelagic swimmers capable of long-range movements that other rays do not engage in. Furthermore, to help in feeding, mobulid rays have also developed conspicuous appendages placed on both sides of their head, called cephalic fins, that can be unfurled and help to funnel their tiny planktonic prey into the mouth. Rolled up tight when the rays are not feeding, cephalic fins are reminiscent of horns, which explains the name "devil ray": another silly common name unfairly bestowed on these gentlest marine creatures.

Most devil rays like to live in tropical latitudes. Still, spinetail devil rays are an exception, being also tolerant of more temperate waters, making them the only ones within their family that not only are found in the whole of the circumtropical belt, but also in the Mediterranean. They are outstandingly beautiful, coal grey on the back, with a conspicuous black "collar" on the nape and bright white on the ventral side. They can reach a width of 3.5 m and weigh over 300 kg. As their name implies, spinetail devil rays are the only devil ray species that has conserved a small, albeit harmless spine at the base of its tail. Their global status is assessed as "Endangered" in the Red List of Threatened Species because they are fished unsustainably in many parts of the world and often end up drowning when accidentally caught in nets, even without being the target of that particular fishery. These fishes' ability to replace mortality losses is very reduced because females produce only one large pup at a time, and not even every year. Spinetail devil rays are endangered in the Mediterranean to a great extent because they die when accidentally entangled in illegal driftnets. Nowhere in the Mediterranean are these rays targeted by a directed fishery, except in Gaza.

---

[16] The spinetail devil ray, *Mobula mobular*, is assessed as Endangered in the Mediterranean: https://tinyurl.com/mrypm5an.

Back to the news of the dozens of devil rays "stranded" in Gaza in 2014: as it turned out, the rays were in fact captured in purse seiners by local fishers, something that happens consistently during late winter and early spring, in some years up to 370 of them being captured in a single month. The news hit the media in 2014 by chance, and went viral for mysterious reasons. The international spotlights having been turned on the Gaza beaches, for once not on the subject of political turmoil, the event had aroused Mohammed's concern, and his investigations created the opportunity for the two of us to come in contact and establish a collaborative relationship.[17] I am still waiting to meet Mohammed in person: the blockade erected around Gaza has always prevented this from happening.

Information gleaned from the fishing activities in Gaza revealed that spinetail devil rays occur off the southeast Mediterranean coasts only between the end of winter and the beginning of spring. Although catches are for the sustenance of the Gazans, considering that animal protein is hard to come by in the country, and despite the relatively low number caught—a few hundred per year out of a population that is certainly at least in the tens of thousands—there is concern that this source of mortality might be unsustainable if combined with all other threats these animals are subjected to. In search of solutions and to understand better aspects of the ecology, population size, and migratory behaviour of spinetail devil rays in the Mediterranean, Mohammed, myself, and other colleagues have joined efforts and embarked on a satellite tagging experiment. The project is ongoing, but the first results indicate that the devil rays undertake a yearly migration, bringing them to the western Mediterranean in summer and back to the eastern Levantine Sea in winter, where the waters are the warmest they can find in the Mediterranean in that season. A connection between such movement and the need to alternate their presence in feeding and breeding grounds is also a hypothesis, but it is too early to tell at this stage.

That being the case, I knew the chances of sighting any of these rays here during summer were next to zero. Still, I also knew that such chances would increase substantially as I moved to the western portion of the Mediterranean in the coming days.

---

[17] Abudaya M., Ulman A., Salah J., Fernando D., Wor C., Notarbartolo di Sciara G. 2017. Speak of the devil ray (*Mobula mobular*) fishery in Gaza. Reviews in Fish Biology and Fisheries 28:229–239. https://doi.org/10.1007/s11160-017-9491-0.

As I approached the coast of Egypt, the navigation had begun to demand more attention because the maritime traffic had become increasingly intense. Getting closer to the Port Said area, I was getting nearer to a shipping lane thick with vessels of all sorts, including large container ships and tankers, carrying their goods to and from the Mediterranean across the Suez Canal. This was no time to entrust the steering wheel to the autopilot, and it was a good thing to be navigating these waters during daylight, as it was by the time I was nearing my destination.

The waters here were beginning to lose the deep blue colour of the typical Mediterranean open seas to acquire a more greenish hue indicative of higher concentrations of phytoplankton, facilitated by the fertilising effect from the greatest of all African rivers. Soon the tall, white lighthouse of Ras al Barr appeared on the horizon, indicating the location of the mouth of the Damietta branch of the Nile. Docks and buildings on both sides flank the waterway's opening to the sea, which serves as a harbour for fishing boats. I had decided to land there instead of in Damietta's commercial port further to the west, and it was here that I made my entrance in the late afternoon.

As soon as docking was completed, and having jumped on the pier, it was a strange and exciting sensation for me to set foot on African soil. However, the excitement was more conceptual than physical, given that the place looked very much like many other Mediterranean places. The dock I had tied *Pontoporia* to merged onto a nicely paved promenade leading to the tip of the pier and the imposing Ras al Barr lighthouse. Crowds of families were strolling along to enjoy the cooling breeze gently blowing from the sea, some pushing a pram with their baby, and kids zooming back and forth with their small scooters. But the link with the great continent looming somewhere to the south was made tangible by the presence of the River Nile that had reached its final destination in that very spot—the way rivers typically end without ever ending. Conspicuously located at the base of the promenade was a tall, shiny black stone with the following inscriptions in Arabic and English on the top of a stylised map of the Nile:

In the name of Allah.
Most gracious. Most merciful.
He has let free two seas Meeting together:
Between them is a Barrier which they do not transgress.
Ar. Rahman (19–20).
Here, The Long Journey of 6695 km.
of the River Nile has ended.

Day 21: Damietta, Egypt

The search for diesel to top up *Pontoporia*'s fuel tank, which started my day in Damietta, turned out futile as I finally discovered, after having gone on a wild goose chase, that none was available in town. However, I was assured that I would have been able to find all the fuel I wanted in Alexandria, where I would stop anyway along my westward course. Luckily, my fuel reserve, helped by the minimal consumption of *Pontoporia*'s little engine, was sufficient to reach my next destination.

After a short siesta to build up energy for the upcoming nocturnal navigation, I exited Damietta's harbour in the early afternoon to hug the coast of the Nile Delta. I had to give a wide berth to the line of land here to avoid getting *Pontoporia*'s keel stuck in the muddy bottom. The sediments discharged into the sea by the great river had caused the waters to be dangerously shallow quite far offshore.

Of all the morphological features of the Earth's surface, rivers are the ones that most strike human imagination as being alive because they are entities in constant movement from where they are born to the sea, where their journey ends. But it is only their journey that ends, not the rivers, so the rivers never die unless they dry up. Consequently, it is natural to think about rivers as entities possessing a life of their own and even some personality. This is not such a crazy notion. Take, for instance, the Whanganui River in New Zealand, which was granted in 2017 the same legal rights as a human being due to the local Māori tribe's fight in the courts to recognise the river as their ancestor. The move went well beyond its romantic connotation and had practical implications, as it created a legal parallel between harming the tribe and harming the river, and managed to leverage much greater protection than would have been possible through a more customary treatment of the river as an inanimate entity to be owned, managed, or protected.

Alas, no such feat is likely to benefit the Nile, one of history's most revered rivers, almost 6700 km long and with a drainage basin affecting 11 countries, its presence having benefited scores of peoples across millennia and, most famously, the ancient civilisation of the Pharaohs. "Egypt was the gift of the Nile", wrote the Greek historian Herodotus, and for good reason. It is hard to think of a greater gift for the peoples coping with the harshness of the Sahara, the world's greatest desert, than the provision of the constant and unlimited supply of fresh water the Nile assures. Unfortunately, modern Egyptians have taken advantage of this resource in quite different ways by damming the river—with the Aswan High Dam being the most recent and impacting example. Indeed, economic benefits were reaped by the works in stabilising floods and generating hydroelectric power. However, the Nile was never the same after the Aswan High Dam was completed in 1970. Nor were the same the amounts of sediments and nutrients contributed by the Nile to the Mediterranean Sea, which caused a sensible increase in the erosion of the coastline surrounding the Delta and, most importantly, had a catastrophic effect on marine fisheries, with catches in sardines and shrimps reduced to a paltry percentage of pre-dam levels because of the decreased influx of nutrients.[18] Needless to say, reports of damages mainly concentrated on the socio-economic effects of reduced fisheries caused by the Aswan High Dam, glossing over the disruption of the broader marine food web with much wider consequences than the narrow human interests involved.

The cascading effects of damming the Nile are certainly not an isolated case of humans playing God with the planet's natural mechanisms and processes without worrying about the consequences of such tampering, intended or otherwise. Giving such practices the pompous name of "geoengineering" makes these acts of hubris even more disturbing. The disruption of climate caused by the human-made accumulation of greenhouse gases in the atmosphere—itself a significant example of unwitting manipulation of our life support systems—has already been calling for extravagant initiatives such as injecting aerosol in the atmosphere to block sunlight, thereby triggering who knows what other cascading effects—at the same time bolstering the alibi for the continuation of belching carbon into the atmosphere. Besides, screening the planet from solar radiation would do little to prevent seawater from becoming more acidic as atmospheric CO_2 dissolves in the ocean, another emerging threat to marine life.

When Ferdinand de Lesseps and his associates got busy in 1858 with the plan of cutting a canal across the Egyptian desert, creating a waterway to connect the Atlantic and the Indian Oceans without forcing ships to

---

[18] Aleem, A.A. 1972. Effect of river outflow management on marine life. Marine Biology 15:200–208. https://doi.org/10.1007/BF00383550.

circumnavigate the whole of Africa, they likely had no idea of the biological effects of artificially connecting the separate Mediterranean and Red Sea ecological realms. They probably did not consider that the Canal might have opened the door to a multitude of marine organisms from the Red Sea and broader tropical Indian Ocean that have since invaded the Mediterranean and come in it to stay, having undertaken a specific type of migration called "Lessepsian" in homage to the resourceful French diplomat. This is what has happened, and it has been much more than just an unwitting ecological experiment on a massive scale. Whilst waters have become increasingly warmer due to climate disruption, tropical invaders large and small have found in the Mediterranean a very favourable environment with few or no natural predators, and their presence is now having significant effects on the ecosystems they have invaded, including the displacement of native species.

Since the opening of the Canal in 1869, more than 400 organisms native to the Red Sea have been identified in the Mediterranean Sea, including fishes, crustaceans, molluscs, echinoderms, and jellyfish, and probably many others still unidentified. Fishes alone have numbered 188 species at the present time, more than half of which are firmly established and happily reproducing in their new residence.[19] These included fearsome species such as the devil firefish,[20] also known as lionfish, a predator able to cause the decline of many other fish species it can prey upon, and the silver-cheeked toadfish,[21] known to cause damage to fishing nets, and also deadly poisonous if consumed by humans; as well as avid herbivorous species from the Siganid family, commonly called 'rabbitfishes' which I had already discussed when visiting the Dodecanese, and that have been responsible for significant changes in sublittoral algal biodiversity throughout the eastern Mediterranean. Admittedly, not all the tropical fish species that have invaded the Mediterranean recently have a Red Sea origin. Many have come in from the tropical Atlantic through the Strait of Gibraltar, taking advantage of the warming of the oceans caused by global change. However, most of these invaders—around 70%—have entered the region from the Red Sea through the Suez Canal. The damage is mainly ecological, but not only, as it has led to considerable economic and human health issues. Take, for instance, the swarms of the Red Sea nomad jellyfish[22] that have caused significant damage to bathers across mid-eastern

---

[19] Golani D., Azzurro E., Dulcic J., Massuti E., Orsi-Relini L. 2021. Atlas of exotic fishes in the Mediterranean Sea - second edition (F. Briand, Ed.). CIESM Publishers, Paris, Monaco. 365 p.

[20] The devil firefish *Pterois miles*, is also known as common lionfish, https://tinyurl.com/3huhrnxk.

[21] The silver-cheeked toadfish, *Lagocephalus sceleratus*: https://tinyurl.com/j2w43spe.

[22] The nomad jellyfish, *Rhopilema nomadica*, https://tinyurl.com/3vnk3m97.

beaches, clogged fishing nets, as well as water intakes of desalination and power plants in Israel.[23]

If Lessepsian migration of organisms from the Red Sea into the Mediterranean has been and still is the most relevant, bringing about significant ecological changes on the receiving end of their travels, not the same can be said about movements in the opposite direction, from the Mediterranean to the Red Sea, known as anti-Lessepsian migration. This is because the predominant flow along the canal is from south to north, which induces organisms travelling in their planktonic larval phase to go north instead of otherwise. This circumstance has reduced invasions from the north and favoured those from the south. However, some Mediterranean species capable of moving out of their own volition, such as fishes, have been found on the Suez side of the canal and beyond.[24] It is generally believed that contrary to the Mediterranean for Red Sea species, the Red Sea is an unsuitable environment for most Mediterranean species, mainly because of the higher temperature of its waters and the abundance of local species. One problem with understanding the actual dimension of the anti-Lessepsian migration is that a north-to-south migration is easier to go unnoticed than a migration in the opposite direction. For instance, when dolphins such as the Indian Ocean humpback dolphin are observed in the Mediterranean, it is easy to understand where they came from because they are a typical Red Sea species. By contrast, when a common bottlenose dolphin is observed in the Red Sea, in-depth investigations such as DNA analyses are necessary to determine whether it had come from the Mediterranean because the species is regularly present on both sides of the canal.

If Monsieur de Lesseps may be forgiven for not having imagined such eco-disasters in the mid-1800s, not the same can be said about the leaders of contemporary Egypt. I don't think they could benefit from such benevolent understanding for acts they implemented a century and a half later, particularly considering that options for mitigating Red Sea invasions into the Mediterranean didn't seem to be unavailable. Consider, for instance, the Bitter Lakes—natural expanses of water found in the southern tract of the Canal, so-called because of their very high levels of salinity. In the early years of the

---

[23] Galil B.S., Boero F., Campbell M.L., Carlton J.T., Cook E., Fraschetti S., Gollasch S., Hewitt C.L., Jelmert A., Macpherson E., Marchini A., McKenzie C., Minchin D., Occhipinti-Ambrogi A., Ojaveer H., Olenin S., Piraino S., Ruiz G.M. 2015. 'Double trouble': the expansion of the Suez Canal and marine bioinvasions in the Mediterranean Sea. Biological Invasions 17:973–976. https://doi.org/10.1007/s10530-014-0778-y.

[24] Golani D. Fricke R. 2018. Checklist of the Red Sea fishes with delineation of the Gulf of Suez, Gulf of Aqaba, endemism and Lessepsian migrants. Zootaxa 4509(1):1–215.

existence of the Canal, the Bitter Lakes worked as a barrier to the Red Sea invasions because of the salinity gradient, which most organisms could not cross. With time, this salinity gradient was reduced by water flow across the Canal. However, the final coup de grace was dealt by Egypt's decision in 2014 to significantly increase the Canal width to allow for the simultaneous transit of ships in both directions. If, before 2014, invasive species already had a wide-open door to the Mediterranean, what they now have is a highway. The Canal is, of course, a formidable economic resource for Egypt (the revenues in 2021 were in the order of US $6.3 billion) and also significantly facilitates trade between the East and West sides of the world. However, other compounding circumstances should have been considered, such as the warming of Mediterranean waters induced by climate disruption and the reduced inflow of fresh water and nutrient-rich silt from the Nile into the eastern Mediterranean caused by the Aswan High Dam; both factors ended up making conditions in the eastern Mediterranean even more like those of the Red Sea. I would have thought it more elegant on the part of the Egyptian leadership to at least try to do something sensible, such as rebuilding a salinity barrier by directing into the Bitter Lakes either a modest flow of fresh water from the Nile or the brine resulting from the surrounding desalination plants.[25]

Regardless of the problems deriving from invading species in the Mediterranean Sea, climate disruption is indeed making things worse in the region, where it is already visibly acting. A consistent warming trend has been observed in the Mediterranean Sea surface temperature over the past 40 years, with values in the Levantine Sea having reached an increase of over 1.7 °C.[26] Lo and behold, the effects of such change did not take long to become apparent, including an increasing frequency of heat waves, the drying out of land due to less precipitation, and the previously unheard-of unleashing of Mediterranean hurricanes.

[25] Peleg O., Guy-Haim T. 2019. 150-year-old idea to limit Suez invasions. Nature 575:287.

[26] Pisano A., Marullo S., Artale V., Falcini F., Yang C., Leonelli F.E., Santoleri R., Buongiorno Nardelli B. 2020. New evidence of Mediterranean climate change and variability from sea surface temperature observations. Remote Sensing 12,132. https://doi.org/10.3390/rs12010132.

The night was dark, save for a moon at its last quarter, and a slight warm, damp breeze blew from land, just enough to fill the sails and push me along on my course towards Alexandria. On the port side, over the near coast, a glow of the sky beyond the horizon revealed humanity's presence farther to the south, in my imagination all the way to the centre of Africa. But it was another type of light that materialised around me as the darkness of the night had descended and plunged me in a state of rapture for the beauty I was suddenly engulfed in. *Pontoporia* was leaving a shiny trail in her wake, where her hull had whipped up with her progress multitudes of light-emitting planktonic microorganisms.

There is little mystery left in marine biology textbooks by the phenomenon of bioluminescence. Once called by sailors "sea fire", bioluminescence is caused by the presence near the sea surface of multitudes of tiny single-celled planktonic organisms, appropriately called *Noctiluca scintillans*, that have the property of transforming chemical energy into light when mechanically stimulated. What still defies understanding is the function of the phenomenon. Some evidence exists that these tiny beings might be less prone to being eaten up by other slightly bigger beings when flaring up; so, a predator avoidance strategy likely justifies the extra energetic cost of the production of such luminous flashes by the scintillating *Noctiluca*. This is possible. However, as I keep remarking when considering the phenomena of life that appear to the eyes, in no way the ability of science to explain observations by revealing the underlying biochemical, physiological, or behavioural mechanisms can diminish our parallel disposition to marvel at such displays of perfection and beauty. If anything, the sense of awe that pervades us is increased, not decreased, by our understanding of how they work and what might have been the evolutionary reason for such traits to evolve.

That night, the diffused twinkling by *Noctiluca* in the dark waters caused by the advance of *Pontoporia* was mixing with the reflection of the stars on the sea surface in an eery combination of live and inanimate agency, which was mesmerising. And then, all of a sudden, a dozen torpedo-like shapes, their skins glowing with the same luminescence, came from nowhere at great speed and positioned themselves around the boat for a few minutes, accompanying her stride like a squad of ghostly escorts. The visitors were bottlenose dolphins, without any doubt, who happened to be in the surroundings and, having heard the approach of *Pontoporia* had decided to make a quick visit to play in her wake. Their visit lasted a very short time but remained etched in my retina for a whilst and in my memory forever. I wondered how many times I had similar visits during my night crossing, which I had been unaware of because the water had no sea fire in it.

Day 22: Alexandria, Egypt

Shortly after daybreak, the entrance to the port of Alexandria came into view in front of *Pontoporia*'s bow, but this time, it was not like entering any port. I was now coming in what might have been the largest port of Mediterranean antiquity, built between the mainland and the island of Pharos, upon which a monumental lighthouse had been made during the reign of Ptolemy II Philadelphus during the third-century BCE.

At about 100 m high, the Pharos lighthouse, long destroyed since, was one of the tallest human-made artefacts of the time and classified as one of the Seven Wonders of the Ancient World. This was known as the world's first lighthouse, conceived and built to help mariners find their way to a safe harbour through the light from a fire built at night on the top of the tower. Such an innovative and helpful idea made Alexandria the world's most modern and welcoming port of the times. The name of the island on which this first lighthouse was built—Pharos—was soon extended by antonomasia to all lighthouses in Greek and all the Latin-derived languages. Modern lighthouses, of which there are now more than 18,600 worldwide,[27] can flash with different colours and periodicities that allow the mariners to tell nearby lighthouses apart and understand where they are sailing in the dark of the night. How many times, back in the days when satellite-assisted navigation had not come of age yet, have I found myself timing the flashes of the various lighthouses on the dark coast in front of me to identify them on the map and derive from that knowledge my position to steer the boat to safe harbour—at times not without some anxiety because the weather was rough and there were dangers ahead. Today, all this is a thing of the past. The GPS allows one to determine one's position with incredible accuracy by measuring through small, cheap, and now ubiquitous contraptions the difference in time of arrival of signals from a constellation of artificial satellites. This would have been pure science fiction just a few decades ago.

Entering the port through the narrow opening between the breakwaters, I hauled to starboard alongside the western pier, past the imposing Citadel that Mamluk Sultan Qaitbay had built in the fifteenth century to protect Alexandria from the Crusaders, using masonry from the ruined lighthouse. I docked in front of the Yacht Club of Egypt as muezzins were calling for prayer from the minarets of the many mosques scattered across the neighbourhood,

---

[27] World's lighthouses: https://tinyurl.com/5bhu34ax.

sending across the still air of the morning that evocative melody which never fails to send a shiver through the spine regardless of whether one is a believer or not.

Having entered with high expectations such an iconic epicentre of Mediterranean civilisation, I was surprised to get a strong sense that Alexandria was, more than any other city I knew, like a chameleon that keeps changing her skin colour century after century. Founded in 331 BCE by Alexander the Great, the city grew rapidly, becoming Egypt's capital under the reign of the Ptolemaic pharaohs and soon the largest city of antiquity before Rome took over the primate. Alexandria was the leading intellectual and cultural centre of Hellenistic times and was the seat of what was thought to be the richest library of the ancient world. After becoming a major centre of early Christianity, the city had lost most of its historic relevance even before the Muslim conquest in 641 CE. Still, it again became a major centre of the international shipping and trade industry in the nineteenth and early twentieth centuries thanks to her strategic placement on the land separating the Mediterranean and the West from the Red Sea and the Orient. Trade had transformed Alexandria into a melting pot of cultures, with a mix of Arab, Ottoman, Greek, and Jewish communities, the quintessence of the Levant, which the city had become the banner of. In the twentieth century, Alexandria's genius loci had permeated Lawrence Durrell's classic tetralogy, the "Alexandria Quartet", weaving together the stories of various Alexandrian socialites and diplomats through events in the city before and during World War II.

Alas, once more, I did not have the time to abandon *Pontoporia* for the time needed, a few days at least, to verify how much of all of the above was still there. At a very superficial level, the feeling was that the city had changed her skin once more, except for the din of the traffic, the confusion, the smells, and the dust that probably have been part of the city's features since time immemorial. All I could do was secure the essentials to restock *Pontoporia's* store-room with supplies and the fuel I needed to jump across the sea that separates Egypt from Crete, my next destination.

However, I could not resist taking the time to sample Alexandrian cuisine before leaving Egypt. About 30 min walking distance from the dock, I found a well-known restaurant where I enjoyed a practical demonstration of the virtues of foul, i.e. stewed fava beans joined with tahini, a sesame paste; falafel as crunchy as they should be (of which I secured a decent supply to carry back aboard *Pontoporia*); shakshuka (eggs cooked in tomato sauce); and topping it all with Om Ali (meaning "Ali's mother"), a delightful sort of warm bread pudding covered with nuts and dried fruit.

The way ahead of me was long and required at least 60 h of navigation to reach the next destination, and it was time to detach myself from the African continent, which I had just brushed upon. A decent breeze had developed in the afternoon, blowing from the sea to the heated land, which allowed *Pontoporia* to sail in the eye of the wind to follow a direct course towards Crete, 400 miles away. By the time the dark had descended, and the breeze subsided, the lights of the land had already disappeared from view, and the Alexandrine humanity only appeared from the glow projected by the city towards the dusty sky. I continued in the night, proceeding under *Pontoporia*'s engine's push in the company of its lulling drone.

Day 23–24: Cretan Sea

The next day, I continued advancing westward towards Crete in a sea known as the "Herodotus Basin", which appeared quite empty of any form of life. The calm surface, only rippled in patches by capillary waves under the soft breath of air moving erratically, provided optimal conditions for spotting any animal that happened to break it. Still, nothing appeared to relieve my sense of solitude. This part of the Mediterranean is, in fact, the one that looks like the emptiest of marine life based on maps showing sightings of cetaceans, and no Important Marine Mammal Areas nor Important Bird and Biodiversity Areas had been identified here. However, this is also a portion of the Mediterranean where scientists have not been particularly busy searching for marine megafauna. One of the commonest conundrums afflicting marine science, and more broadly ecological science, is deciding whether the lack of data indicates a lack of observers or a lack of study subjects. Ideally, the problem is addressed by increasing the presence of observers—something scientists call "effort"—across the study area as much as possible and then expressing the number of observations in relative rather than absolute terms, e.g. the number of sightings of a given species per kilometre of effort. The resulting encounter rates are then comparable across different spaces and times. But even this trick does not solve the problem entirely, given the habit of marine megafauna of existing at relatively low densities in most places, like in the Mediterranean, which decreases the probability of making a sighting. A combination of low density of animals and low effort—like here in the Levantine Sea, where very few sighting cruises have occurred until now—can result in a zero-encounter rate, belying the presence of animals. Add to the mix that the significance of results one can obtain from such observations depends on the density of the data collected, which means that the fewer observations, the higher the uncertainty about what one can deduct from such observations. Seasonality can also be a factor, with marine animals moving around the Mediterranean depending on the time of year, like spinetail devil rays, for example at times in search of richer feeding grounds or otherwise to reach locations more suitable for breeding. Unfortunately, ecological knowledge of Mediterranean megafauna until now, particularly in its eastern portion, has been collected mainly during summer, when weather conditions are more favourable for researchers and being at sea is safer, especially with smaller craft. The result of all this is that the picture we have of the distribution of many whale, dolphin, and shark species is likely to be quite different from the real one. Hopefully, the application of innovative technologies—e.g. through satellite imagery or investigations of the presence of animals through analyses of their DNA left in the water—will eventually provide answers to such basic ecological questions and lay the grounds for more solid knowledge bases.

Scanning the horizon whilst wondering whether nothing would appear, either due to real emptiness or because of an artefact caused by poor sampling power, eventually conveys to the observer a sense of inevitability: a condition that, in turn, leads to an attention drop, thus compounding the problem. Such a state of affairs brought me to proceed in the company of my thoughts throughout the day and the following uneventful night. And so it was after the sun rose again in *Pontoporia*'s wake, chasing the dampness of the night and warming my back. Later in the morning, a gentle breeze started to be felt as I was getting nearer the dynamic influence of the Cretan landmass, still well beyond the horizon to the north, being warmed by the rising sun.

I was approximately 60 miles from my destination, proceeding under sail, when I again entered the Hellenic Trench IMMA, a large portion of sea running in a broad arch from the Ionian Sea south of the Peloponnese and Crete, all the way beyond Rhodes to the east. I had briefly crossed the area's eastern end when leaving Rhodes 10 days before, and I was now entering it again on the western side after having gone full circle clockwise around the Levantine Sea.

This area was identified as important because it is the home of the two foremost deep divers in the Mediterranean: the Cuvier's beaked whale (I had encountered a mother and calf pair after leaving Rhodes), and the mighty sperm whale.[28] The expectation of encountering sperm whales certainly was one of the primary reasons for me to embark on this journey because sperm whales rank amongst the oceans' most extraordinary creatures for many reasons, and their presence in the Mediterranean contributes to making this sea a special place. Sperm whales are extreme in many ways. With males on average 16 m long and weighing 45 tonnes, they are the largest predators living on the planet today: a size they can afford to maintain mostly because they subsist on mid-water and deep-water squids, a virtually untapped source of food. To catch their squid, however, sperm whales had to develop extreme diving capabilities to hunt the squishy critters at depths generally between 500 and 1000 m, and often beyond. Their thick skull accommodates the largest brain that ever existed in a living being, about 8 kg on average in adult males. What they do with such impressive neural hardware is a mystery, given our ignorance of their cognitive abilities. All we can say is that sperm whales live in societies made of complex networks spread across the oceans over thousands of miles. One can guess that large brains might help to manage, at best, such social complexity, made even more challenging by the vast distances involved. The basis of sperm whale society is the family unit, typically up to a dozen individuals, composed of adult females and their young. When the young males grow, they leave their mothers and sisters to form groups of bachelors. Eventually, as they become older, they end up living in solitude except when briefly joining family units to mate (Fig. 6.4).

In the vast underwater expanses in which they roam, sperm whales stay connected through their vocalisations, which they produce with a monumental nasal complex—up to about one-third of the whale's length—resting within their heads above the upper jaw's bone. The organ is encased in a sac containing a special wax called spermaceti, which owns this silly name to the old whalers' belief that the sac contains the animal's sperm (never mind that the whalers seemingly failed to wonder why females also had large quantities of "sperm" in their heads). The sounds produced through this complicated nose, serving both a social function and as a sonar to locate prey in the darkness of the abyss, are loud, broadband clicks resembling the beating of a

---

[28] The sperm whale, *Physeter macrocephalus*, is assessed as Endangered in the Mediterranean: https://tinyurl.com/jwpyrnku.

Fig. 6.4 A large male sperm whale, swimming at the surface seconds before diving, shows its powerful body, broad flukes and asymmetric, left-pointing blowhole (Illustration by Massimo Demma)

hammer on a hard surface. These were the sounds made by mysterious "carpenter fishes" in the imagination of the puzzled US Navy operators, the first in history to listen to underwater sounds when engaged in detecting enemy submarines during World War II.

With very few exceptions, such as the Red, Black and Baltic seas, and the Persian Gulf, sperm whales are found just about everywhere in the world's oceans, wherever the waters are deep enough to host their prey. The Mediterranean population, thought to be genetically distinct from sperm whales living in the Atlantic Ocean, is present in greater numbers in the western part of the basin probably because there is more food there. Still, a nucleus of about 250 survives in Greek waters, concentrating in particular along the Hellenic Trench.[29] The existing knowledge of sperm whales living in the eastern Mediterranean and their plight was collected almost singlehandedly by friend and colleague Alexandros Frantzis, a Greek scientist whom I had already mentioned for his research in the Gulf of Corinth, who has dedicated his life to the Mediterranean representatives of this most intriguing character. In his gripping addiction to sperm whales, I think that Alexandros could be described as the antithesis of Moby Dick's Captain Ahab, where the engine of his

---

[29] Rendell L., Frantzis A. 2016. Mediterranean sperm whales, *Physeter macrocephalus*: the precarious state of a lost tribe. In: G. Notarbartolo di Sciara, M. Podestà, B.E. Curry (Editors), Mediterranean marine mammal ecology and conservation. Advances in Marine Biology 75:37–74. https://doi.org/10.1016/bs.amb.2016.08.001.

connection to the whale is fed by love instead of hatred. For decades, summer after summer, Alexandros Frantzis has cruised the Hellenic Trench back and forth searching for the whales, approaching them, spying on them gently, painstakingly piecing together bits of information on their lives like and archaeologist reassembling the scattered tesserae of an incomplete mosaic, and sending philippics against the military, industrial and shipping constituencies that have been endangering the whales with their actions.

The notion that sperm whales occur in Greek waters dates to the earliest years of the Western world's zoology, as their existence there was well known to Aristotle, who wrote about them 2370 years ago with a description that leaves no doubt about the fact that he knew what he was writing about. Alas, Aristotle himself, in his superb discernment, could not have foreseen that Mediterranean sperm whales, particularly the ones living in Greek waters, would eventually become an endangered tribe as denounced by Alexandros Frantzis: under siege by humans who are surrounding them with lethal (and illegal) driftnets in which they drown, hitting them with the bows of ships that travel too fast to allow the whales to get out of the way, and expelling them from their habitat with noise produced to search for oil in the sea bottom[30]—in a day and age in which nations are purportedly resolving to leave such crude where it is, untouched, to stem the increase of emissions of CO_2 in the atmosphere.

As maritime traffic in the Mediterranean keeps increasing, one of the main hazards that large cetaceans such as fin and sperm whales have to deal with is the risk of being hit by a ship.[31] Being air-breathing mammals, whales are forced to spend time at the surface to breathe; for example when they sleep. This brings them into the danger zone where they can be killed or, at a minimum, injured by the bow or propeller of a vessel. One could suppose that the noise a ship generates as it advances would be enough to warn the whale of her approach and cause it to get out of the way. Still, things are not that simple considering how many whales are sighted bearing on their bodies the deep scars derived from a too-close encounter with a vessel. A recent study convincingly concluded that maritime traffic presents

---

[30] Notarbartolo di Sciara G. 2014. Sperm whales, *Physeter macrocephalus*, in the Mediterranean Sea: a summary of status, threats, and conservation recommendations. Aquatic Conservation: Marine and Freshwater Ecosystems 24 (Suppl. 1):4–10. https://doi.org/10.1002/aqc.2457.

[31] Panigada S., Pesante G., Zanardelli M., Capoulade F., Gannier A., Weinrich M.T. 2006. Mediterranean fin whales at risk from fatal ship strikes. Marine Pollution Bulletin 52:1287–1298.

a hazard to the survival of Mediterranean whales, with a particular risk incurred by fin whales.[32]

I will never forget the sighting we made in 2005 in the Ligurian Sea when a large whale—a fin whale, in fact—lifted its tail out of the water to dive right in front of us, and in doing so, we could all see with dismay that half of the tail was gone—almost certainly, sliced off by a ship propeller. The wound appeared to be perfectly healed, and the whale—whom we had nicknamed on that occasion "Mezza-coda" ("Half-tail"), had no signs of having health issues of sorts. Only the behaviour of lifting the tail out of water when starting a dive—something fin whales do not do, unlike sperm whales—indicated that the whale had to change its diving technique by giving that extra push down-wards to compensate for the decreased thrust forward a half-tail could afford.

Unfortunately, it does not seem that much can be done systematically to address the problem of ships hitting whales in the Mediterranean. The most critical factor at the root of the problem is the vessel's speed, dramatically increasing the chances of colliding with a whale above the rate of 10–12 knots. Such speed limitation would not be cripplingly incompatible with cargo and tanker traffic. Still, reducing speed is considered very hard to accept by ferry companies that compete with each other mainly by bringing their passengers from one side to the other of their passages in the shortest possible time, and therefore, adopt speeds that are usually above 20 knots. However, mitigation actions can be envisaged on a local scale, mainly under the auspices of the International Maritime Organisation (IMO), the agency of the United Nations in charge of regulating international shipping, including human safety at sea and addressing the effects of maritime traffic on the marine envi-ronment. We will see examples of this when cruising in the Western Mediterranean.

Protecting sperm whales frequenting the Hellenic Trench from the risk of being hit by ships has been one of Alexandros Frantzis' main concerns for years. The area is crossed by intense maritime traffic consisting of vessels con-necting the western Mediterranean with the Aegean, particularly the port of Piraeus and, further up, the Dardanelles, Istanbul, and the Black Sea. However, the area is also essential for the life of sperm whales living in the eastern Mediterranean. After years of lobbying by Alexandros, in February 2021, finally, the Greek Navy issued a notice informing mariners about the presence of the whales in the area. One would think that institutional initiative could

[32] Sébe M., David L., Dhermain F., Gourguet S., Madon B., Ody D., Panigada S., Peltier H., Pendleton L. 2023. Estimating the impact of ship strikes on the Mediterranean fin whale subpopulation. Ocean & Coastal Management 237, 106,485. https://doi.org/10.1016/j.ocecoaman.2023.106485.

have gone further than that, but at least this was a timid signal that governmental bodies were taking notice of the issue.

As is increasingly the case in these strange times, the best answer to governmental pussyfooting comes from the industry itself. So it was that in January 2022, the powerful MSC Group, a leading international shipping and logistics conglomerate, unilaterally decided to reroute its ships along the west and south coasts of Greece with minor but significant changes to reduce the risk of collision with sperm whales in the area. The company adopted this measure after a coalition of advocacy organisations, including OceanCare, WWF Greece, and the International Fund for Animal Welfare, had engaged with it based on the ecological data collected by Frantzis' research organisation.

The breeze had died out, and *Pontoporia* was puttering forward in a sea so flat that it looked like I was navigating on oil. The sighting conditions were excellent, and the excitement about the likelihood of running into sperm whales was intense. Still, having proceeded into the heart of the Hellenic Trench IMMA for a couple of hours, I was beginning to lose hope of encountering the dear giants. And then, straight ahead, tiny oblique sprays of vapour appeared just below the horizon, probably a couple of miles away—brightly illuminated by the sun that was beginning to decline to the west. Sperm whales were in sight! Minutes later, at least five silhouettes of the whales' tails flickered above water. The whales had dived all together in unison, and they could be expected to stay down for up to about 1 h, but I had latched onto them now, and I knew that when they resurfaced, I would have been close.

When I arrived on the spot where I had seen the whales diving, I killed the engine and stayed motionless and silent, waiting for them to come up again. The water depth there was almost 1800 m, over a steep slope leading up to Crete's continental shelf, about 20 miles away: perfect squid hunting ground. The silence was so thick that I could hear the thudding of my heart. When the first whale surfaced again and emitted its first blow, it sounded like a gunshot and startled me.

Awe is appropriate to describe the sentiment that one feels when encountering the whale, and the sperm whale in particular. It is a mix of wonder, astonishment at seeing something so big, admiration, reverence, and respect—and, surprisingly, a complete lack of fear. These emotions have nothing to do with the learned, science-infused knowledge of why sperm whales are special, which eventually kicks in later. Nor do they have much to do with the sense of guilt for us as a species, being such a horrible source of trouble for these spectacular animals. It is just an unmediated, almost biochemical animal-to-animal feel, the consideration of the whale's immenseness, of this being so big that you can see it only in part because the rest is hidden under the water that swirls around its body in small vortices as it slowly and solemnly moves at the surface.

But it is not just bulk the source of the sense of awe we feel when we find ourselves in front of a sperm whale. It is the sensation of what is hidden within that bulk, a glimpse of which we can get by looking the whale in the eye on the rare occasions in which that eye becomes fleetingly visible from above the water. At the risk of oversimplifying matters to explain myself better, let us compare an encounter with a sperm whale with a terrestrial mammal. Take, for instance, that graceful, wonderful denizen of the woods, the deer. When you watch a deer browsing in a meadow, you feel that you understand that deer because perhaps there is not much to be understood about it. The deer is hungry, the deer is alert, the deer is afraid, the deer is running away. Despite the deer being a reasonably complex and evolved organism, the connection between its behaviour and its state of mind is not so hidden. But when you are in front of a sperm whale—or in front of an elephant, for that matter—you sense that there is a mind behind that eye that you are quite unable to fathom: a reasoning being very likely intent in asking about you that same question that you are asking about him: what is he thinking? Unfortunately, this is, for now, an unanswerable question and one that taints one's excitement with a pinch of frustration: because we are not given the gift or knowing what a sperm whale—the owner of the largest brain that has ever lodged into anyone's cranium—might be thinking about you. It is like spending a whole dinner in the company of a famous savant but with the obligation for both to observe absolute silence throughout the meal.

One by one, a few minutes apart, the whales all emerged: five medium-sized individuals, about 10–12 m long, probably females because two were closely accompanied by smaller ones; and a huge one, over 15 m long; the first who surfaced, definitely an adult male. Eight whales in total, a typical family unit, but with the addition of the large male. This was one of the surprising observations made by Alexandros. In the eastern Mediterranean, large males can be associated in the long term with family units, a departure from the norm observed throughout the world's oceans.

Probably tired after this last foraging dive, the whales were now floating at the surface in loose formation, slowly moving about and resembling more floating logs than living beings if it weren't for the occasional blows. *Pontoporia* was just as motionless as the whales. As I was drifting, the whales got so near that I could observe them in the smallest detail, with the sounds of their breathing surrounding me in a sort of odd, priceless intimacy. It was impossible in this magic moment to avoid feeling a strong sense of community with their intelligence, of shame for having made their lives so difficult, and of appreciation for having been allowed to be in their company for a few memorable minutes.

Reluctantly, it was time for me to leave the whales and resume my way towards Crete. I was concerned that starting *Pontoporia*'s engine might have disturbed the whales, but there was no sign of that. So, I was allowed to slowly put some distance between the boat and the floating congregation and bid them goodbye. One hour later, the Island of Gavdos was abeam, 5 miles to the port side, and then the landmass of Crete, my destination, became visible to starboard. After about 3 h, as the sky was getting dark and the first planets appeared in the sky, I was entering the harbour of Paleochora, delighted to be in Greece again.

7

Southern Mediterranean: From Paleochora to Carloforte

Illustration by Massimo Demma

G. Notarbartolo di Sciara, *Sailing Across a Wounded Sea*,
https://doi.org/10.1007/978-3-031-54597-9_7

Day 25: Paleochora, Greece

Sitting in *Pontoporia*'s cockpit with a cup of coffee as the sun rose over Crete's mountains, the attraction of that wild land was hard to resist. Besides, the many weeks spent at sea had bred in me a craving for setting foot on solid rock and filling my lungs with the dry air of summer, scented by the aromatic essences emanating from the Cretan *phrygana*. The fresh memory of having come in sight of the coast yesterday, hour after hour as I was approaching the end of my passage from Egypt, had left an impression. The land's colour had been changing from a pale blue, barely discernible from the sky above it, to darker shades as I got closer, with the blue gradually morphing into an inviting mix of greens, yellows, and browns with all the grooves indicative of the mountain's asperities. I felt this strong wish, almost a physical need to forge into those mountains, explore the valleys, subject all those impressions to a reality test, and capture the land into my own experience. *Pontoporia* seemed happy to be left to rest in the quiet of the Paleochora harbour for the day, and I set off in exploration along a footpath that from the town's eastern outskirts was leading to Sougia, a remote coastal hamlet 13 km to the east.

Running flat just inland from the pebble beaches of Anidri and Gialiscari, sparsely populated by a humanity intent on roasting its mozzarella-white flesh under the unforgiving sun, the path soon left the welcome shade offered by patches of tamarisks to climb up the steep slope of the coast beyond. The views from the rising path were breathtaking, with the deep blue expanse separating Crete from Cyrenaica on the African side, shimmering under the rising sun; I imagined the waters being populated by those same frolicking sperm whales I had encountered the day before. The path was winding up and down as I progressed eastward, alternating easy climbs to descents to small, secluded coves, until I came to the ruins of Lissos, in the old days a trading hub of some importance, said to have been abandoned after an earthquake destroyed it. Just uphill of the main ruins were the remains of the Asclepeion, a healing temple sacred to Asclepius, the first physician-demigod in Greek mythology. One of about 300 discovered across Greece and a place where pilgrims converged to seek healing, the Asclepeion here was surviving with huge, squared blocks of stone surrounding a floor partially covered with the delicate remains of a mosaic. The place still conserved an aura of serenity despite being reduced to ruins by the vagaries of time, its stillness in fact in itself a mosaic of soft voices perfectly integrated into a soothing soundscape—the goat bells ringing from the uphill meadows, the chirp of cicadas, the murmuring of the breeze blowing across the pine needles, and the gurgling noise of a nearby spring flowing out of a stone spout under a gnarled olive tree.

These sounds surrounded me like a song of sirens inviting the pilgrim to stop, rest in the cool shade of the trees, and heal from whatever evil spirit might trouble the body or mind.

I chose instead not to listen to the sirens and continued my hike. The trail winded uphill out of Lissos around brushes and rocks until it got me into a small valley leading to a pine-flanked gorge overhung by a majestic red cliff ornated with grey and white vertical streaks populated by screaming swifts. And then down again to descend on Sougia, ending behind the town's small harbour. From there, quite conveniently, I boarded a small ferry that took me back to Paleochora, with the bonus of taking yet another look at the coast from a different angle of sea level, the folds of the mountain dramatically enhanced by the lengthening shadows of the dying day.

Day 26: Paleochora, Greece

It was time to resume my journey, and *Pontoporia* and I were both ready and eager to put some more miles under our belts. And a mouthful of them was what I had on my plate for my next stop; the island of Lampedusa in the Strait of Sicily, right across from Tunisia's coast, was a good 540 miles away. If all went well at *Pontoporia*'s pace, it would have taken me at least 3 days to cover that distance. With this leg, I would move from the eastern to the central Mediterranean and be ready to peep around Sicily's westernmost tip into the western portion of the Great Sea. Three days and three nights non-stop were going to be the longest passage since the beginning of the journey, but the boat and I, by now, were an experienced team, and I was looking forward to the big jump. The sea I was going to sail in was exceptionally trafficked—one of the world's most trafficked—with the main ship lane connecting Port Said to Gibraltar passing across the Strait separating Sicily from Tunisia and Europe from Africa. But *Pontoporia*, endowed as she was with all the technology I needed to ensure that the passage was safe, was an intelligent entity.

The fair weather I had benefited from in the Levantine Sea generated by a regime of high atmospheric pressure was now changing, with a medium-strength weather system transiting over the European continent from the Atlantic, creating the conditions for a north-westerly wind in my area. Not a very strong wind, to be fair, something around Beaufort 4, generating a bit of swell and a few breakers here and there. The direction from which the wind was blowing was slightly more from the north than from the northwest, allowing *Pontoporia* to keep her course to Lampedusa under sail, precisely due west, full and by.

Having left Paleochora early in the morning, after about 1 h, I had come out of the lee created by Crete, and the sea was moderately rough. Tilted to port, *Pontoporia* was cutting through the waves with style and nonchalance under conditions where she could give her best. She was making way at speeds with peaks of nine knots, two more than she could have made under engine in a flat sea, and the expectation of arriving at Lampedusa earlier than expected, thanks to the increase in average speed, was a pleasant one.

Although the prospect of progressing under sail was also a pleasant one, the choppy seas in which I was moving forward carried the cost of cutting down dramatically the possibility of sighting wildlife, except for the ubiquitous Scopoli's shearwaters that were accompanying me day and night with their elegant aerial dances over the waves' crest. Nevertheless, small herds of striped dolphins visited me three times on that first day in the Ionian Sea. With their fast, energetic swimming, frequent jumps and splashes, it was hard to miss them when they approached and came to place themselves briefly in front of the bow. They seemed so joyful in their antics, but I knew this was their way of moving, particularly in rough weather, with no particular relation to their mood. Even their smiling faces with their mouths upturned at the corners was just how their bodies were made. Still, not only is the dolphins' sight always an opportunity for natural contemplation and a moment of sheer pleasure for the sailor, but it also gives the sense of being welcome in an environment otherwise rather alien to human nature.

Days 27, 28: Southern Ionian Sea

At dawn, I had already put 190 miles between Crete and me, and I was making good along my course over the deep waters of the Ionian Sea. Conditions had remained similar to the day before. *Pontoporia* was doing all the work, with me assuming the role of a watchful passenger. The wind had not changed since I left Greece, neither in strength nor direction, so I did not even have to trim the sails. All I had to do was supervise that all was all right and perk up when receiving from the radar the alarm signal that the occasional ship had come within a ten-mile radius. Otherwise, I had descended into a state of serene calmness, morphing into a sort of ataraxic automaton, going about the things that a person needs to do to survive—eat something, drink something, take a nap regardless of whether it is day or night, read a book, fix something that needs fixing (and god only knows how many things need fixing on a boat, always)—watching the parallel slipping by of the hours and the miles under *Pontoporia*'s hull without the minimal expectation for time to pass faster, or yearning to reach the destination. It is a beautiful state that rarely happens in everyday life on land but is typical of life on a boat on longish passages and, I suppose, perhaps also of astronauts travelling in space. Free to wander with the mind at will without having anything impellent to worry about, it is that time to explore new mental territories and make new imaginative connections.

One such mental territory kept coming up as I considered the water expanse to the south and what lay beyond. From the day I left Crete until reaching Lampedusa, the whole extent of the coast of Libya was streaming by to my port side. It remained unseen because it was too far (140 miles at its closest point) but quite present both in reality and in my mind. Libya, with her harsh deserts extending for a 1000 km of sands, rocks, mounts, and unliveable arid plains into the heart of Africa. But also with a coastline in places unexpectedly green with plant life, like in Cyrenaica, where hundreds of hectares of soil deposited during millennia hide the still largely undiscovered archaeological treasures left there by important ancient Greek and Roman dwellings. I remember visiting the magnificent ruins of both Leptis Magna and Cyrene in the years before Libya had become a war zone, and having been most impressed by the real-size sculpture of the prow of a ship with a marble dolphin swimming underneath it, like dolphins had done since the beginning of human navigation, and still do to this day under *Pontoporia's* bow.

Before I was born, my father had also sailed regularly in these same waters for some time, but in different and direr circumstances. A commander in the Italian navy during World War II, he was assigned to escort convoys of ships shuttling between Italy and the Libyan lands that were at the time occupied by Italy. It was a dangerous and thankless task to try to protect the helpless merchant vessels under constant attack by the Allied forces that were enjoying a crippling technological superiority, and it was pure chance that he and his crew survived the ordeal, having been forced by the unfortunate geopolitical circumstances to fight in so calamitous a conflict. Once more, I was brought to consider my luck, first for being born to him thanks to the dear man's survival beyond the war years, and second for being able to sail across the same waves without worrying about being torpedoed by some enemy.

At the break of dawn the following day, I had sailed for over 370 miles since leaving Paleochora and reached a spot in the southern Ionian Sea where the waters were deepest along my track, more than 4000 m below the hull. Had *Pontoporia* been a bathyscaphe instead of a sailing boat, at the speed I was going, it would have taken us almost 20 min to reach the sea bottom on a vertical dive—and I would have given an arm to be able to make such an exploration and see what the seafloor was like at that depth.

Soon, however, I was going to cross the "Malta Escarpment", where the bottom rises with a pretty steep slope towards the shallow continental shelf, marking the entrance to the Strait of Sicily. Had the sea level been lower only by about a couple of hundred metres, the land projection across the Mediterranean made by the Italian peninsula and Sicily would have continued southwards, creating an almost complete bridge connecting Europe to Africa and dividing the Mediterranean into two distinct seas. This was the case during major glaciations when large quantities of water were sequestered on land as ice. Under the current conditions, the sea passage between the west and east Mediterranean is uninterrupted. Still, the unseen underwater transition from the deep Ionian Sea is quite abrupt as one approaches that steep seafloor slope.

These configurations of the seafloor are almost invariably exciting places to explore because they act as concentrators of marine life. Surface waters are usually poor in nutrients—most notably nitrates and phosphates—because the photosynthetic microalgae living there use them up when they grow; by contrast, nutrients accumulate in deep waters as a product of the decay of the organisms that have died and sunk to the bottom. This condition creates a disconnect between the microalgae—the primary producers that stay near the surface because they need sunlight to thrive through photosynthesis—and the availability of the nutrients themselves that tend instead to accumulate in the dark of the deep seafloor, unused. As a result, the vast ocean expanses are, to a large extent, barren of life, except for the large animals capable of autonomous movement and that can go a long time without feeding; for example the whales. However, when a horizontal current is pushed against an underwater slope, the water mass is forced to ascend, and the upward motion—oceanographers call it upwelling—brings the nutrient-rich deep waters to the surface. Similar to oases in a desert—where the factor limiting life is water—oceanic oases can occur where special conditions, such as upwellings, annul the disconnect between nutrients and sunlight, one of the main factors limiting life in the open seas. The Malta Escarpment I was crossing on that day was a grand slope, probably the most dramatic seafloor rise in the Mediterranean, with a depth abruptly rising from about -3300 m to a paltry -270 m in less

Fig. 7.1 A round, beakless head, a body so covered with white scratches to become white in places, and a very tall dorsal fin ensure that Risso's dolphins cannot be confused with any other cetacean when sighted at sea (Illustration by Massimo Demma)

than 40 miles. It was time to open my eyes well and expect to run into something interesting (Fig. 7.1).

And, in fact, a whole bunch of fins, distinctly taller than in the dolphins I had encountered until then, came into view at some point between one wave crest and the other, and the strangest cetaceans soon surrounded *Pontoporia*. I wondered when I would finally run into a pod of Risso's dolphins,[1] and here they were. Named after the seventeenth century's Niçois naturalist Antoine Risso, who provided Georges Cuvier with a skeleton needed for the species' formal description, these intriguing cetaceans also go by the name of "grampus", derived from the Latin "grandis piscis", or perhaps from the French "grand poisson".

Whilst it is often not easy for the non-specialist to tell apart the different dolphin species encountered at sea, this is certainly not true for Risso's dolphins. Their identity is immediately evident from many morphological characteristics, such as their body size, which is larger and stockier than the more conventional dolphins, and their big, rounded head conspicuously devoid of the typical dolphin beak. But what is most distinctive is their colouration. Uniform grey at birth, as the dolphins grow older they become covered with scratches all over their bodies, which they receive during social interactions with their schoolmates. Body scratching is commonplace amongst dolphins of all species as an effect of social interactions, playful or otherwise, whereby marks on one dolphin's delicate skin are left by the raking from another's teeth. However, in other species, these marks rapidly heal and, if superficial as

---

[1] The Mediterranean subpopulation of Risso's dolphin, *Grampus griseus*, is currently assessed as Endangered: https://tinyurl.com/5bvj5vzx.

most of them are, typically disappear in a matter of days. This is not the case in Risso's dolphins, where these scars turn white and remain forever, accumulating with time. The older the dolphin is, the more scratched it becomes. And so it happens that with increasing age, older individuals eventually become outstandingly immaculate all over their backs and around their heads. Why this happens is a mystery, but indeed, these large, old, snowy-coloured animals, clearly visible even when they swim at some depth, have something venerable about them. Their whiteness could signify status to command awe and respect, not unlike silverback gorillas and white-haired wise humans of both genders.

When the tall fins appeared far over the waves, I already had a hint that Risso's dolphins came to inspect *Pontoporia* and her passenger. But any doubt had dissolved once the white, well-built torpedoes appeared through the translucent walls of the waves, their light backs reflecting a spectacularly bright aquamarine hue as they took position all around the boat. This was precisely what Risso's dolphins usually do with vessels: they are often attracted by them, probably out of curiosity, which brings them to approach the boat. Unlike smaller dolphin species, however, Risso's are not too keen bowriders, a pastime they rarely engage in. They instead prefer to circle or swim alongside the boat, attentively observing the moves of the humans aboard.

Typical haunters of slope areas, the cosmopolitan Risso's dolphins are found throughout the Mediterranean. Still, their numbers seem to be decreasing, possibly due to bycatch in fishing gear, chemical pollution, plastic ingestion, and human-made noise. Due to these circumstances, the Mediterranean population, which is genetically separate from that of the Atlantic, is listed as "Endangered" in the Red List. Risso's dolphins in the rest of the world are generally doing better than their Mediterranean relatives. One of them, "Pelorus Jack" as he was known to sailors, used to be quite famous.[2] He (or she?) became known around the turn of the nineteenth century for escorting vessels across the shoals of the Cook Strait between the main islands of New Zealand. He was always seen alone, engaged in this odd routine of associating himself with vessels for more than 24 years. The appearance of Pelorus Jack was a welcome sight for mariners because the dolphin apparently was guiding the ship across the passage, which was made dangerous by strong currents and shoals, a merit which (unsurprisingly) did not prevent some hoodlum from trying to kill him.

I turned on the hydrophone to hear if the dolphins had to say anything, aware that the ambient noise generated by the waves and by *Pontoporia*'s hull

---

[2] Pelorus Jack: https://tinyurl.com/4bf4798b.

riding them might have masked the dolphins' signals. I should not have worried, for the dolphins were vocalising so loudly that I would have been able to hear them in a gale. Risso's dolphins have a very funny vocal repertoire, with a tendency to emit trains of crackling sounds that can mutate into utterances that resemble the most irreverent of raspberries.

The Risso's dolphins who approached *Pontoporia* as I was coming into the Strait of Sicily probably did not think that I needed to be escorted along my way because after having circled the boat for a few more minutes, they all moved to a spot about a 100-metres away and started lingering about lazily. The behaviour was interesting, so I stopped the boat's momentum by putting the bow to the wind and backing the jib, to see what was happening momentarily. At one point, one dolphin's tail was seen protruding vertically from the water and coming up by at least 1 m, the animal performing an underwater headstand. The sight was bizarre by itself, but it was just the beginning. Seconds later, another tail came up, and then another and another. Soon, there were at least six or seven tall tailstocks with the beautifully shaped cetacean tails at their top, slowly waving in the air: quite something to be seen in the middle of nowhere, this strangest animal behaviour resembling a modern art installation of some crazy sculptor. Cetacean tails, with their smooth curved lines, are so elegant to watch: they have something that we cannot avoid considering inherently beautiful, like the curves of a violin, a letter in Arabic, or the shape of a woman's body. Head-standing by Risso's dolphins is well-known, as they often exhibit such spectacular yet utterly mysterious propensity. But these spectacles are short-lived. Minutes after they started the show, all the dolphins resumed their more conventional horizontal posture and moved on. After that, the school disappeared into the depths, and the animals were gone for good.

Just before sunset, I was passing south of the main island of Malta, somewhere up north about 20 miles to starboard. Given the increased maritime traffic, as the moonless night came upon me again, it was time to keep my eyes well open. All the ships longitudinally crossing the Mediterranean during their long-distance passages had to funnel through the narrows between Sicily and Tunisia. The seas were beginning to be dangerously busy in the area.

Day 29: Strait of Sicily

I safely arrived in Lampedusa early in the morning and docked *Pontoporia* in the town's spacious, well-protected harbour southeast of the island. Lampedusa, the largest of the Pelagie Archipelago and part of the Italian "Isole Pelagie"

MPA, is under Italy's jurisdiction and is entirely Italian in all respects except for its geography and geological aspect. The island is a piece of Africa thrown over the sea, a slab of Saharan-looking rock lying over the Tunisian shelf 120 miles south of Sicily and 135 miles southeast of Tunis. You would be forgiven for expecting to see a camel caravan treading along the island's dry landscape instead of observing the typical features of a southern Italian town.

Being in sleep debt, as soon as the docking was completed, I lost myself in a long, dreamless slumber. Eventually, though, around lunchtime, appetite started to prevail over the wish to rest, and still half asleep, I found myself sitting at a table of a tavern overlooking the harbour, so close to the dock that I could watch *Pontoporia* gently swaying in her moorings. Meanwhile, a smoking plate of spaghetti al pomodoro was placed under my nose as by magic by the gentlest of hands. Few Italians can survive for more than a handful of days without a meal of pasta. I, of course, was fully autonomous in this respect: cooking pasta was one of the first things I learned from my mother, and *Pontoporia*'s galley was equipped with all the necessary state-of-the-art implements. However, no matter how hard I tried, in no way could I beat the flavour of a sauce made with Lampedusa's tomatoes grown under the African sun and—I could almost say—cooked with the lymph still flowing through their cells, seasoned with basil leaves so fragrant to make your head spin, the optional clove of garlic, but rigorously without grated cheese on top. I could eat spaghetti al pomodoro every day I have left to live in this world—provided they are cooked al dente—without even thinking of getting tired of them.

From my vantage point on the trattoria's terrace, I could see how the port of Lampedusa had changed since the last time I had been there several summers before, when it was simply a place where vacationers were intermingling with the local types of fisherfolk and hotel and restaurant managers getting about their businesses. There was, then, a sort of light-hearted, joyful atmosphere that was now gone, leaving room for a more sombre, ominous, overhanging mood; as a consequence, the pleasure of munching on the spaghetti I had deftly rolled around my fork had all but disappeared. Over the years, Lampedusa had become the entrance gate to Europe for tens of thousands of desperate people running from their homes throughout Africa and Asia and looking for salvation and refuge. A beacon of hope that had attracted to their deaths multitudes, women and children included, thereby transforming the Mediterranean into a boundless graveyard. And the lucky ones who had been able to make it here alive, relieved to have survived through the most horrible ordeals, had to discover that a second ordeal had just begun for them, having landed onto a selfish new world that was not willing to welcome them and tried to callously reject them as much as it could. The thought of all that suffering was acting like the steel jaws of a vice clamping on my chest.

That humans often behave as a migratory species is a biological fact. Our species once migrated out of Africa, where it originated, but we don't have to go so far back in our history to observe human migrations. There were significant migrations in the twentieth century from Europe to the Americas and Australia, and another major one, an unstoppable one, I would add, is unfolding now under our eyes as governance, economic, and environmental debacles conjure to force people away from where they were born. Unfortunately, with eight billion humans on the planet and counting, a perception is shared, as widely as inappropriately, that we have run out of space to accommodate newcomers. Everywhere, the old occupants object to sharing their land, food, and jobs with migrants, fearful that these foreign masses will chip away at their privileges. Dealing with the contemporary migrant crisis certainly is a huge challenge, but one that could be much better addressed if it were not polluted by selfishness, indifference to suffering, and ignorance: all sentiments that are imbibing those same Western societies that are, at least in part, responsible for creating in many developing countries the conditions for people to want to run away. With such a dreadful lack of compassion by humans towards humans, I wondered what hope we can have for a sense of stewardship for the marine environment and its non-human inhabitants. Meanwhile, absorbed in these painful thoughts, the spaghetti had become cold and no longer so appetising. Still, that was an irrelevant price to pay in exchange for my incredibly privileged and meritless condition of being a citizen from a wealthy part of the world.

Day 30: Lampedusa, Italy

Before the sun had risen too high in the sky, I went on a hike from the port along the only road leading to the west of the small island, my body casting a very dark shadow on the ground as the heat of the day was increasing. I was heading for a cove in the south coast, lined by a beautiful beach called "Spiaggia dei Conigli" ("Beach of the Rabbits"). Some wild rabbits indeed inhabit Lampedusa. However, the beach's name apparently had nothing to do with the furry critters, possibly a simple misnomer.

The walk was an easy and pleasant one. Lampedusa is about 10 km long from end to end, and the Spiaggia dei Conigli is only a fraction of that from the harbour. The hike took me slightly more than 1 h as the road climbed gently from sea level to the top of the island's plateau, a paltry 100 m altitude.

The landscape consisted of arid scrubland and looked really like being in the middle of an African desert if it were not for the deep blue marine expanse surrounding me from all directions.

There was a reason for me wishing to engage in this promenade, aside from the usual pleasure of walking and exploring a visited land. Many years before, I had come to Lampedusa briefly, attracted by the fact that the island was then the only site in Italy where marine turtles nested. My visit turned out to be an experience laden with strong emotions (Fig. 7.2).

Marine turtles are peculiar in their breeding behaviour in that they spend their entire life at sea after birth, except for females when they lay their eggs. For this purpose, mama turtles select sandy beaches where they haul out well up beyond the spray zone, preferably at night. Once in the desired location, they start digging a hole in the sand with their hind legs and eventually lay their small, white, round eggs inside the hole. Many of them, up to 120 per nest, are deposited in a process lasting 1–3 hours before the turtle returns to the water, a time in which she is completely defenceless and quite vulnerable to predation—or simply disturbance. After a couple of weeks, the turtle is ready to lay eggs again, for up to six or even more times in a single season. In those early years, more than two decades before this trip, out of the totality of

Fig. 7.2 Surprised at the sea surface, this young loggerhead turtle took her time to look at the people who were watching her before diving away (Illustration by Massimo Demma)

the Italian coasts, loggerhead turtles nested in that single location in Lampedusa's Spiaggia dei Conigli. This made the beach unique, not only because it was considered one of Italy's most beautiful beaches.

Understandably, there was not much of a honeymoon between marine turtles and the Lampedusans at that time. The unique conservation value of the site clashed with local economic interests, one of many examples of conflict between nature conservation and human development. With its fine white sand, the Spiaggia dei Conigli is the largest and most attractive of the island's many beaches. Consequently, it is the most valuable asset for a local economy primarily based on tourism. Setting aside the beach from tourist visitation to avoid disturbing the turtle nests was out of the question. To be fair to the Lampedusans, the conflict here never degraded into a bitter war between the two factions, as it had instead flared up with ugly acts of violence from the developers' side in the early 1990s in Laganas Bay, a significant Mediterranean loggerhead nesting site on the Greek island in Zakynthos. Here in Lampedusa, the conflict was eventually solved through compromise. Turtle protectionists were allowed to carry out their activities on site intermingled with bathers. Nest locations were carefully marked from time zero early in the summer, fenced off with small round wire enclosures to avoid inadvertent trampling by the beachgoers. There, nest sites were lovingly monitored throughout incubation until the eggs hatched a couple of months later. Tourists, made aware of the situation by the presence of the motley turtlers' crew, were, of course, accommodating and, most of the time, intrigued as well.

In recent years, things have hugely improved for loggerhead turtles in the Mediterranean, particularly in Italy: a resounding conservation success story and a testimony to the effectiveness of marine turtle conservation efforts in the region. Despite the large number of deaths of loggerhead turtles that are accidentally ensnared in longlines used to catch swordfishes and tunas, their status in the Mediterranean was changed in 2015 from "Endangered" to "Least Concern" in IUCN's Red List,[3] which is an extraordinary result. Today, loggerheads nest regularly in several sites in Sicily and Sardinia and occasionally in up to 70 other locations disseminated along the Italian peninsular coasts—quite an improvement from the not-too-distant days when nesting occurred in a single place here in Lampedusa.

---

[3] The Mediterranean subpopulation of the loggerhead turtle, *Caretta caretta*, is currently assessed as "Least Concern": https://tinyurl.com/bdehfhj3.

As I walked down the trail connecting the main road to the beach, the memories of that early visit came back forcefully. We were already at the beginning of autumn, and further to the north, on the continent, the weather had already broken with rainy days and a sensible drop in temperature. In Lampedusa, it looked instead like summer was never going to go away, and the beach was full of people intent on catching the year's last sunrays like living solar panels. That year, only a single turtle nest existed, fenced off near the beach's east end. The eggs had been laid there at the beginning of August, and hatching was due at any time, as three turtle volunteers in attendance explained. The team was led by Mauro Cocco, then a doctoral student at the University of Roma "La Sapienza". I spent the best part of the afternoon lazily resting on the warm sand near the turtle nest, chatting with Mauro and his associates and being brought up to speed about the conservation status of Mediterranean marine turtles. At dinner time, I returned to the hotel for a shower and a quick bite. The plan was to get back to the beach and spend time there—hoping in the luck of witnessing the happy event … That night, there was going to be a full moon. Experts deny the existence of any significant connection between the moon phase and the moment in which turtle eggs hatch; nevertheless, I was looking forward to spending time in good company on the sand, still warm from the day, surrounded by the beauty of a calm, moonlit night.

Coming back along the trail leading down to the beach brightly lit by the moon, some commotion around the nest area had become visible, with the volunteers moving around with their torchlights flashing right and left, and hushed utterances. And indeed, when I rushed to the nest site, the first hatchlings were starting to come out from the ground. It was a magic spectacle, and I still get goosebumps when I think of it. There was this multitude of tiny beings, perfect copies of a loggerhead turtle scaled down to a few centimetres in length, pouring out with dogged energy from the ground in their life's first opportunity for being in motion. From a condition of serene, almost contemplative quiet, there suddenly was mayhem on the beach, with the few bystanders who were on site at that hour gathering in excitement and with the volunteers clearing the way for the procession of little creatures scuttling with determination towards the water's edge, plunging head-on into the wavelets and disappearing into the dark sea. For a few minutes, the beach buzzed with animation and muffled cheers as the onlookers watched such an unexpected and moving show of natural exuberance. A handful of minutes indeed, and then, as the last tiny turtle had gained the sea and disappeared into the water, there was silence again. The whole event had been so fast that it was like we had all had a collective dream. It reminded me of a sleepy train station out somewhere in the boondocks, where a flurry of excitement arises when a train arrives, with the departing passengers coming out as if from nowhere and the arriving ones descending from the carriages, and then, only minutes after, everybody has gone their ways and silence reigns again. The shiver of emotions I experienced from witnessing the scene was a combination of excitement for the manifestation of life and concern for the fate of the defenceless little things that now had to swim in the dark waters, full of predators, minutes after having come out of their shells. Based on robust statistics, I can guess that very few of them will have survived that first night of their post-hatching life.

Back to my second visit to the Spiaggia dei Conigli, a sense of déjà-vu was inevitable, with the same beach populated by a new generation of vacationers, but very similar to their forebears, and a similar little group of volunteers getting organised, perhaps the children of the ones I had met before: determined to save a species with their actions, which in the meantime had already been saved through the dedication of those who had come before them. But mind you, loggerheads in the Mediterranean are still far from being safe. Certainly, conservation actions on breeding sites have been very effective, and a change in popular sentiment towards marine turtles has helped quite a bit. But a more insidious threat—global warming—is lurching at the horizon. The sex of turtle embryos, as they develop inside their egg throughout the incubation time, is determined by the temperature of the sand they are laid in, which must optimally be comprised between 28 °C and 32 °C.[4] This temperature range would ensure a balance of the sexes of the hatchlings. Above the highest limit all hatchlings will end up female, and male below the lowest. Until now, the situation was balanced, but as the Mediterranean summers get hotter like they have been recently, we can expect an increasing female-to-male ratio in the population, with potentially disruptive consequences that not even the tenderest of care at the breeding sites can prevent. The case of the recovery of loggerhead turtles in the Mediterranean holds a vital conservation lesson, very

---

[4] Yntema C.L., Mrosovsky N. 1982. Critical periods and pivotal temperatures for sexual differentiation in loggerhead sea turtles. Canadian Journal of Zoology 60(5). https://doi.org/10.1139/z82-141.

well described by colleague and friend Randall Reeves in a recent essay he wrote on the conservation of marine mammals: "The task is dynamic and never-ending; our plane of understanding is in constant flux, as is the smorgasbord of threats to the animals. The harder we look, the more problems we uncover. We must keep acting on many fronts and never become complacent if there's to be any hope of keeping these magnificent animals with us in the long term".[5]

Before returning to the port and to *Pontoporia,* I climbed back to the main road. I continued towards the island's west end, north of a deep gully where a pine forest managed to hold its ground against the wind and the area's dryness, until I reached a point along the north coast that afforded a 360° view of the surrounding sea. Here, in front of this coast, was where fin whales, the authentic Mediterranean royalty, congregate during winter to feed. An IMMA had been identified around Lampedusa[6] to recommend respect for the site. But the whales occur here only during winter, whereas in summer they concentrate in the northwest portion of the Mediterranean. That was where I was counting on meeting them as I would get closer to *Pontoporia*'s journey's end.

Right across the water expanse to the west, a mere 70 miles away, laid the coast of Tunisia, another North African nation that evoked many personal memories. After a short flight from my hometown in Milan, I find it always enjoyable to dive into the heart of the Maghreb, a region that is so close to home and, at the same time, so different. Tunis, in particular, despite being considered a somewhat "westernised" Arab city and being the capital of a country having close ties to southern European nations like Italy and France, is the site of a lively market—the kasbah—and its many mosques and chaotic traffic leave no doubt of in which part of the world one is. I had been in Tunis countless times to interact with colleagues working at the Regional Activity Centre for Specially Protected Areas established there by the UN Mediterranean

---

[5] Reeves R.R. 2022. Cetacean conservation and management strategies. Chapter 1, pp. 1–29 in: Notarbartolo di Sciara G., Würsig B. (Eds.), Marine mammals: the evolving human factor. Springer Nature, Gewerbestrasse 11, 6330 Cham, Switzerland. 465 p. https://doi.org/10.1007/978-3-030-98100-6_1.

[6] Lampedusa IMMA: https://tinyurl.com/4w7ytdv9.

Action Plan. There have been meetings to promote Mediterranean MPAs and discuss strategies to protect Mediterranean-threatened species, including cetaceans, sharks, rays, and the monk seal. One person stands out as a constant in my interactions with the Centre and within the broader collaboration framework to protect the Mediterranean marine and coastal environment across the widest range of nationalities. Chedly Rais is, by training, a marine ecologist who has been involved for most of his lifetime in creating connections between science and policy across the often tricky cultural, social, and economic differences that span the Mediterranean region from north to south and from east to west. Working with Chedly for decades, during which we developed a deep friendship, I had many opportunities to appreciate how someone who has received scientific training can double as a policy expert and even, on occasion, a fine and very patient diplomat. Of all my travels to Tunisia, I remember one with particular fondness. Chedly and I had often entertained the conviction that a wise redirection of fishing efforts to cater for tourists interested in the marine environment—beyond the experience of collecting sun rays on a beach—could have significant conservation effects. Artisanal, small-scale fishers involved in such practice, dubbed "pescaturismo" ("fishery-tourism") in Italy, where the initiative had started, could manage to earn their livelihood with less hardship than the one deriving from everyday fishing practice; a share of the tourist public would become educated on the subject of making a living at sea; and impact of fishing on the marine environment could decrease. We decided it would have been good to promote that idea in the Kerkennah Islands, a small group of low-lying islands not far from the coast of southern Tunisia, home to a community of fishers engaged in traditional activities. So, in December 2012, I travelled to Tunis with Davide Petrini, an Italian fisherman/entrepreneur who has spearheaded "pescaturismo" in Liguria, and we were met at the airport by Chedly, who drove us to the port of Sfax. From there, we boarded a ferry to reach the islands.

The Kerkennahs look like shards of Sahara, with their complement of date palms and camels, thrown onto the Mediterranean—just like Lampedusa. The people of the islands have developed a unique traditional technique of passive fishing called "charfias", very simple systems of weirs based on converging arrays of palm fronds stuck in the sandy seabed, placed in such a way as to channel the fishes swimming in the ebbing tide into a capture chamber lined with a net. Traditionally, charfias are installed only during the winter months to avoid interfering with the fishes' breeding season, and are rebuilt each year by people who have a deep knowledge of the local currents, seabed contours, and tidal dynamics—and are as sustainable as fishing can be. Before presenting the "pescaturismo" concept to an assembly of local operators, we were taken to inspect one such charfia. We could note that scores of small fishes had indeed entered the capture chamber, where they were kept until they would have been harvested. I could observe how many fish species had been targeted in that particular instance of charfia fishing. This traditional fishing technique was inscribed in 2020 on UNESCO's List of "Intangible Cultural Heritage of Humanity" with all its connected rituals, including community participation, praying, and meal sharing.[7] Ten years after our visit, I was happy to learn that the Kerkennahs have become a destination for international ecotourism. Presumably, however, charfia fishing will not be central to tourism activities because of the seasonal mismatch.

Day 31: Lampedusa, Italy

The onboard barometer indicated a rise in atmospheric pressure, as a typical summer anticyclonic regime was stabilising again over the central Mediterranean. The north-westerly winds that had given me a push from Crete had died down, leaving a remnant of swell that would disappear the next day. I left the harbour of Lampedusa early in the morning under power, hugging her south coast in a westward direction, and as I doubled the western cape of the island, I headed north, towards the west tip of Sicily.

The Strait of Sicily is an interesting place in many respects. Its shallow depths mark the division between the eastern and western Mediterranean. For millennia, the Strait has been a crossroads for peoples busily moving across the basin in all directions to explore, trade, wage war, or migrate. As far as humans are concerned, in modern days, the Strait is crossed from south to north by the world's destitute striving for a safer and better life, and from east to west by a stream of ships loaded with goods for the world's affluent. Today, the Strait is the site of one of the world's busiest shipping traffic lanes: 15% of the global shipping traffic is concentrated within an area that is less than 1% of the world's oceans, and most of it is crammed in the narrow maritime space squeezed between Sicily and the coast of North Africa.

But the Strait of Sicily is not only relevant to human movement. It is also a major bird flyway across the Mediterranean, connecting Europe to Africa along the Italian peninsula and islands, used twice a year in either direction by multitudes of birds despite the efforts of armies of hunters (legal and otherwise) intent on destroying the very survival of the migratory pathway. Below the surface, other nomad tribes regularly cross the Strait under their constant

---

[7] Charfia fishing in the Kerkennah Islands: https://tinyurl.com/5n73rd25.

urge to move: the great bluefin tunas as they push into the central Mediterranean from the west in early summer for their breeding rituals; spinetail devil rays at the same time of year but in the opposite direction to take advantage of the greater availability of food afforded in summer by the western basin; and gigantic fin whales as they move from the Sicily Strait waters towards the northwestern Mediterranean in early spring, attracted by the availability of krill[8] offered by the Ligurian Sea's productive waters. By contrast, sperm whales seem to have an aversion to venturing over the shallow shelf waters of the Strait. To move across the Mediterranean, they prefer to negotiate the narrows of the Strait of Messina between Sicily and continental Italy, which connect the deep Tyrrhenian and Ionian seas where these whales feel so much more comfortable.

The broad shelf area south of the Strait of Sicily, which takes the name of Tunisian Plateau where it extends along the coast of North Africa, is also of fundamental importance for one of the most beleaguered groups of Mediterranean marine species—sharks and rays—the plight of which I had already the opportunity of recalling when I was sailing down the length of the Adriatic Sea. According to experts from the International Union for the Conservation of Nature, at least 53% of the 86 species of sharks and rays residing in the Mediterranean Sea are at risk of extirpation and require urgent action to conserve their populations and habitats.[9] Conservation problems in the Mediterranean are caused to these species almost exclusively by fisheries: either because they are fished at speeds that exceed their replacement rate, which is very low, or because they get accidentally hooked in lines or entangled in nets without being the targets of the fishery. Still, they die in the process and are often discarded. Just as an example, guitarfishes, represented in the Mediterranean by the common guitarfish and the blackchin guitarfish—having a status, respectively, of "Endangered" and "Critically Endangered"—once commonly found throughout the Mediterranean, are now confined to the region's eastern and southern waters, such as in the Tunisian Plateau and off Libya, where however their capture is, quite absurdly, still allowed.

The Plateau contains sites important for reproduction—nursery areas, as they are called—of many shark and ray species. The first and foremost that comes to mind is the white shark, *Carcharodon carcharias*, as the fearsome creature is known to science: not just one of the many inhabitants of the Mediterranean, but one that commands awe and respect at the very mention

---

[8] Krill are small zooplanktonic crustaceans of the order Euphausiacea, most notably a key food item for many large plankton-feeding organisms: https://tinyurl.com/4649vw5u.

[9] IUCN on Mediterranean shark and ray status: https://tinyurl.com/4x3ync7c.

Carcharodon Lamia

Fig. 7.3 I have never had the chance to encounter a white shark in the Mediterranean, nor am I aware of any good photograph having ever been taken of a free-swimming white shark in this sea. I have therefore resorted to reproduce here this fine plate from my collection, part of a work on the Italian fauna (*Iconografia della fauna italica per le quattro classi degli animali vertebrati*) that was published in 30 instalments between 1832 and 1841 by Carlo Luciano Buonaparte, a notable nineteenth-century Italian zoologist and nephew of Napoleon

of its name. The Tunisian Plateau is one of the best-known areas where juvenile white sharks are found and is thought to be a vital nursery area for the species in the Mediterranean (Fig. 7.3).

With a length exceeding 6 m and weighing over 2 tonnes, the white shark is not the largest marine predator. However, we cannot avoid considering it as a more formidable being than, say, the much larger sperm whale because, unlike the whale, we humans happen to be—albeit very rarely—on the shark's menu.

The fear of being eaten is an unforgiving instinct we still have solidly hardwired in our minds, even though humans have managed for a long time to keep their predators at bay. The number of people killed and eaten today by animals, such as crocodiles and some of the largest carnivorous mammals, is insignificant. Even compared to that number, humans killed by sharks are even more insignificant. The white shark in the Mediterranean is a case in point. In the period comprised between the first historically proven event on record (which happened in 1907 in Malta when a white shark attacked a fisherman who had fallen overboard) and the present day, we know of only 54 unprovoked attacks by sharks—primarily, but not exclusively, by white sharks—and a mere 21 of them lethal. The last recorded shark attack in the Mediterranean occurred in Tuscany in 1989.[10] Twenty-one casualties in 115 years, i.e. less than 1 in 5 years, in a sea where the number of potential victims exposing their meaty body parts to shark predation is in the order of the hundreds of millions every summer. Compare that with the 8000 household fatalities which occur on average each year in Italy alone. Or compare it with worldwide statistics informing us about the proportion of humans killed by sharks in comparison to those killed by mosquitoes (140,000 times), by snakes (27,000 times), by hippos (600 times), or even by deer (26 times) and horses (4 times), to realise that the chance of being attacked by a shark in the Mediterranean is practically non-existent. This is made even more apparent when considering real tragedies, such as the loss at sea of 1600 migrants in the Mediterranean in 2021 alone, reported by the UN High Commission for Refugees.[11] It was drowning that killed these desperate people as they crossed aboard unseaworthy watercraft the Strait of Sicily, defined by incompetent media as "shark-infested waters", whilst encounters with sharks have been reported not once by survivors.

How can the presence of the fearsome marine predator be consistent with the scarcity of shark attack incidents in the Mediterranean? The answer is simple: white sharks do not fancy human meat. In the Mediterranean, they are primarily after tuna, but they are also known to grab the occasional dolphin when they manage; once upon a time, when monk seals were more abundant, they too might have been part of the white shark's diet. These fishes are also avid scavengers when they run into floating animal carcasses, including dead whales, with even that rotten flesh seemingly more attractive than human meat.

Despite its harmlessness, the white shark's notoriety as a mean killer is so deep-seated in people's minds that there is no love lost on the broader public when considering the bleak future of the Mediterranean population, the status of which is regarded as "Critically Endangered" in the Red List.[12] Despite being a protected species, the white shark's condition is so dismal that it is not unlikely that its days in the Mediterranean are numbered, particularly if

---

[10] Notarbartolo di Sciara G., Bianchi I. 1998. Guida degli squali e delle razze del Mediterraneo. Franco Muzzio Editore, Padova. 388 p.

[11] 1600 Migrants Lost at Sea in Mediterranean This Year: https://tinyurl.com/3b49mph5.

[12] The white shark, *Carcharodon carcharias*, is assessed in the Mediterranean as Critically Endangered: https://tinyurl.com/5tth7jmh.

officials in countries like Tunisia keep turning a blind eye to illegal and indiscriminate killings in its nursery grounds by unscrupulous fishers.[13]

Despite all the hours I have spent at sea whilst encountering and observing marine life, I have never been able to come face to face with a white shark in the Mediterranean, and I know of very few people who have. And yet, when swimming in open waters and despite the awareness of being safe generated by ecological wisdom, a slight chill invariably runs through my spine at the thought that something big and sharp-toothed might lurk somewhere below, animated by bad intentions. Invariably, however, the idea is shrugged off and pushed to the background of my mind, but there it remains.

At the same time, though, I cannot help feeling a sense of admiration for these survivors. Would the Mediterranean be a safer place without white sharks? Of course not, given the insignificance of their attacks on humans. Instead, the day the last white shark of the Mediterranean will have been killed in a Tunisian net, this sea will be a different place, its sense of wilderness further diminished. Indeed, I relish the thought of white sharks being here and wish that they would never go away, because the Mediterranean is not a chlorinated swimming pool but still a mosaic of vibrant ecosystems providing habitat to a bewildering array of fascinating forms of life—we humans being one of them, and in my humble opinion, not the most interesting one.

But there is another reason Mediterranean white sharks should be protected: they have a very intriguing origin. Genetic comparisons across different populations of white sharks have revealed that the Mediterranean population is more closely related to conspecifics from the South Pacific than to white sharks found today in the North Atlantic, generating the hypothesis that a South Pacific nucleus might have been conveyed during the Pleistocene Era from the Pacific across the Indian Ocean and then around the tip of Africa into the Mediterranean, without coming into contact with their North Atlantic conspecifics.[14] A bold idea but consistent with the available facts, and one more reason for striving to keep the Mediterranean white sharks from falling off the cliff and into oblivion.

---

[13] Milazzo M., Cattano C., Al Mabrouk S.A.A., Giovos I. 2021. Mediterranean sharks and rays need action. Science 371(6527):355–356. https://doi.org/10.1126/science.abg1943.

[14] Gubili C., Bilgin R., Kalkan E., Karhan S.U., Jones C.S., Sims D.W., Kabasakal H., Martin A.P., Noble L.R. 2010. Antipodean white sharks on a Mediterranean walkabout? Historical dispersal leads to genetic discontinuity and an endangered anomalous population. Proceedings of the Royal Society of London Series B Biological Sciences 9 p. https://doi.org/10.1098/rspb.2010.1856.

On my way north after doubling the western tip of Lampedusa, the cliff of Capo Ponente layered with spectacularly coloured slabs of limestone, I set *Pontoporia*'s course towards Sicily past the island of Pantelleria, another small piece of Italy, famous for her passito wines and exquisite capers.

Contemplating the beautiful, deep blue waters I was crossing, it was difficult to avoid considering how such beauty was hiding a habit of abuse. The first thoughts naturally go to the suffering of my kind, the tens of thousands of people who are risking their lives in these waters in their search for a better future. But grief is not limited to just people. We have already seen the predicament white sharks live in throughout the Mediterranean and here in particular. However, they are not the only ones who risk being extirpated from their critical habitat in the Strait of Sicily. Many other species are likely to undergo the same fate.

Take monk seals, for instance, which we have already considered earlier in this journey. These pinnipeds have long disappeared from this area through relentless persecution, but in this case, we hope they might be coming back. Many shark species, by contrast, may not be able to spring back from oblivion. An excellent example of this is the sandbar shark. One of many shark species assessed as Endangered in the Mediterranean,[15] this beautiful, medium-sized, harmless shark congregates seasonally in the waters surrounding the lonely rock of Lampione, a mere 10 miles to the west of Lampedusa, and included in the "Isole Pelagie" MPA. Lampione always had dubious fame amongst sea goers in Italy as "the place where sharks are", purporting a sinister notion of dangerousness for anyone unfortunate enough to be in the surrounding waters for any reason. If there is any place in Italy where the idiotic expression "shark-infested waters" is commonly used, Lampione is undoubtedly one of them, forgetting that sharks occupy the habitat they evolved to live in.

If there is a species that can be described as infesting any of the planet's waters with immensely destructive power, a non-marine species by the way, that species is *Homo sapiens*. Far from being "shark-infested", Lampione is

---

[15] The sandbar shark, *Carcharhinus plumbeus*, is assessed in the Mediterranean as Endangered https://tinyurl.com/mtvrm5s7.

glorified by the seasonal presence of a community of these sharks, as documented by the studies of colleague Marco Milazzo and his team from the University of Palermo.[16] I had never been to Lampione to pay a visit to sandbar sharks and had no time to stop there on this journey. Still, I am aware of other places where females congregate to give birth to their live pups, like, for example in Türkiye, a location very close to the coast north of Bodrum. A few such sites are known in the Mediterranean, although, rather puzzlingly, the shark aggregations in Lampione do not seem to be explained by reproductive behaviour.

After a full day of navigation under engine, in the late afternoon the summit of Pantelleria had finally come into view out of a misty horizon, slightly to the port side of the bow. As I was approaching the island's east side on my way north, my attention was attracted by a minor commotion of the smooth surface a few hundred metres to starboard. Something big was there, and the notion gave me a flush of adrenaline, waking my senses after a full day of having only seen empty waves. I pointed *Pontoporia*'s bow in the direction of whatever it was that had disturbed the surface, turned the engine off, and proceeded with the residual headway in complete silence until I reached the spot. A couple of minutes later, the shape of something big was again near the surface, producing telltale wavelets as it was moving about slowly. And then the animal revealed itself—or rather the animals because what I had seen at first was not alone—when two parallel fins emerged through the surface, as if two sharks swimming side by side in perfect synchrony, and then two more a bit further, and finally a third couple of fins at a greater distance. These fin pairs were slowing down, coasting under the momentum provided by a swim push, which had just ceased, but were maintaining with choreographic perfection their parallel position. No wonder, for each pair of fins did not belong to different individual sharks but were the wingtips of a single spinetail devil ray. I had, by chance, run into a group of three of these most elegant of fishes, moving about on the warm sea surface, who knows with what purpose in

[16] Cattano C., Turco G., Di Lorenzo M., Gristina M., Visconti G., Milazzo M. 2020. Sandbar shark aggregation in the central Mediterranean Sea and potential effects of tourism. Aquatic Conservation: Marine and Freshwater Ecosystems 2021;1–9. https://doi.org/10.1002/aqc.3517.

mind—possibly some type of social interaction, given that they were not engaged in feeding behaviour, and no zooplankton concentration seemed to be in sight at the surface. As they moved away from the setting sun's glare to the west, their bodies became visible across the transparent water, with their grey backs turned to light blue by the water's absorption, in stark contrast with the black "collar" they had over their napes. The rays did not seem at all bothered by the motionless *Pontoporia* and the equally motionless, transfixed bystander. Intent in their dancing, the rays were frolicking around *Pontoporia* initially, but eventually they moved away and were no longer in view. I wished them good luck (driftnets, still unlawfully used to catch swordfish and tuna in Sicily, are deadly traps for these fishes) as I started the engine again and resumed my way to the north. Around sunset, Pantelleria was now close to port, and I hugged its east coast as I continued towards the Egadi islands.

As I proceeded in the dark of the night, I had to be extra careful about maritime traffic because I was crossing a most intense shipping lane. Every day, hundreds of vessels—bulk carriers, oil tankers, container ships, and various other boats—cross the Mediterranean on their way from the Indian Ocean to the Atlantic Ocean or vice versa. In doing so, they all have to be at some point between Pantelleria and Sicily, where I was going to be as well. The distance between Pantelleria and the Egadi Islands was a mere 72 miles, but most of the traffic was quite close to Pantelleria, so I had to pay the greatest attention at the beginning of the night.

I felt like I had to cross on foot a highway with heavy traffic when forging ahead in the dark across the shipping lane, eyes glued to the screen of the AIS receiver and occasionally to that of the radar to make sure I had all the targets in check, as some ships have the bad habit of turning their AIS signal off. Unlike smaller, nimbler vessels such as *Pontoporia,* which can alter course at the last second and slow down their speed to zero in a few tens of metres, large ships should be considered like meteorites, hurdling blindly along their path and unable to alter neither speed nor course for a long time after the initial reaction. In no way can a large ship do anything to avoid ramming you in case you are so inconsiderate to place your boat in front of her bow, even at a considerable distance. Fortunately, most ships navigate at low speeds to save fuel, between 8 and 12 knots, and that night, slaloming between them by passing close to their sterns and far from their bows proved not to be too much of a challenge. I knew that the few ships navigating at speeds greater than 20 knots posed the most significant threat. Still, luckily, none of them were around when I finally exited the main shipping lane as I approached the western tip of Sicily. From then on, the navigation was more relaxed, and I could enjoy the luxury of brief moments of dozing.

Day 32: Strait of Sicily

Early in the morning, the Egadi islands had appeared out of the sky above the hazy horizon, as the west coast of Sicily with the city of Marsala was streaming by to the starboard side. Right here, in these waters, in 241 BCE, the Romans had engaged the Carthaginians with a mighty fleet of 200 warships in what became known as the "Battle of the Egadi (Aegadians)", and by destroying the enemy's fleet, they managed to bring the First Punic War to an end. As a result, Carthage surrendered Sicily to Rome, which led to the beginning of the Roman dominance of the Mediterranean.

None of that drama had survived the millennia as I was approaching the small Egadi Archipelago—Marettimo, Favignana, and Levanzo—in the light of the new day, surrounded by the shimmering of capillary wavelets covering the calm surface. Favignana and Levanzo lie very close to the coast of Sicily, whereas Marettimo, the least inhabited of the trio where I was directed, stands, tall and wild, 17 miles offshore. Favignana is primarily well-known because it used to host in the nineteenth and twentieth centuries what might be considered the Mediterranean's most important factory for processing bluefin tuna. Established in 1874 by the Florio family, with the annexed tuna traps ("tonnare") used to capture the great fishes as they come into the Mediterranean from the Atlantic Ocean on their breeding migration, the magnificent Tonnara of Favignana was turned into a museum by Sicily's regional government as a splendid example of industrial archaeology. The smaller Levanzo lies not far to the north, and is famous for its Palaeolithic and Neolithic rock paintings in the "Cave of the Genovese", including the earliest known Mediterranean art representations of a dolphin and a bluefin tuna. Marettimo glows in her splendid offshore isolation and outstanding natural beauty, and it was there that I decided to head for the night.

As I doubled Punta Bassana at the island's southeastern tip, the tiny fishing hamlet with its small harbour—less than 700 souls—came into view midway along its steep east coast. As I approached the port sailing along the mountainous coast, it was as if the island was broadcasting an almost tangible invitation to explore. Unlike the other two islands closer to the coast of Sicily, tamer in their geology and more contaminated by civilisation, Marettimo has managed to retain her wild soul through physical distance and rough terrain and, consequently, her powerful attractiveness.

Altogether, the whole archipelago is comprehended within the Egadi Islands MPA, established by Italy in 1991 to safeguard its marine and terrestrial natural environment. For management purposes, the area was subdivided into zones having different levels of protection and different regulations, and—quite puzzlingly—the largest of such zones allows fishing with bottom trawling, which is amongst the most destructive and yet legal fishing methods, as I mentioned when I was sailing down the length of the Adriatic. This fact, and many others concerning Italian MPAs, can be understood when considering that the push for creating MPAs in the late twentieth century as a bulwark against the rapid environmental destruction visible everywhere was met with violent opposition by the locals, and compromise had to be accepted. The situation later changed significantly with the passing of years, when it dawned on the local communities, with economies to a large extent based on tourism, that the establishment of MPAs was playing in their favour. But the original fishing privileges enshrined in the compromising legislation could not be changed and are probably there to stay for a very long time, absurd as they are.

The puzzling configuration of the Egadi MPA stimulates a general and rather pitiless consideration of the nature of protected areas. In very general terms, an MPA is a delimited marine space recognised as deserving special attention because of its natural value, and needs to be managed so that the goals of its designation are attained. Even this simple definition underscores some problems and issues. First, the need to establish a dedicated legal regime over a marine space to protect its special natural value is humanity's admission that it cannot have a respectful relationship with our environment without recurring to legal measures. Unlike the other Earthlings, humans are prone to damage the world they live in as a habit, and when they find a space that they think is especially valuable, they need to erect fences around it to protect it against themselves. This is, to me, a rather peculiar notion by itself. Second, once a law is passed by a nation's parliament establishing that an area within that nation's jurisdiction is legally protected, actual protection needs to happen through appropriate management and, therefore, by establishing and funding a management body. Despite being gazetted with all the legal bells and whistles, an MPA is not protected if it is not managed for protection with competence and a clear legal mandate, which requires establishing a management body, hiring professionals, and the consequent commitment to long-term expenditure. Unfortunately, many decision-makers worldwide often ignore this evident platitude and are misled to think about their excellent job of protecting some valuable natural space by simply passing the law to that effect, often without a proper budget. Too many of the existing protected areas in the Mediterranean are still mere "paper parks", as exemplified by assessments performed by the WWF. Although 9.7% of the Mediterranean Sea was designated as protected by the legal systems of the various coastal nations, only 2.5% of such surface is covered by MPAs that are supported by a management plan. But the numbers get worse: the surface covered by MPAs where a management plan is effectively implemented shrinks to 1.27%, whereas the surface covered by MPAs that are fully protected, i.e. where no fishing is allowed, is a paltry 0.03%. In other words, except for a sea surface that can be contained within a little square having a side of 27 km, nowhere in the region have decision makers been able or willing to tell their fishers: "You have the whole Mediterranean to roam about, but please leave in pace the fishes and their habitat within the boundaries of this place".[17] This is only because, unlike the tuna, the swordfishes, the groupers, and the mackerels, fisherfolk can vote. In such a desolating landscape of political ineptness, it might now be easier to understand why it is legally possible to exercise bottom trawling over about half of the Egadi MPA.

I eventually docked in Marettimo's only harbour, surrounded by a couple of dozens of small fishing boats belonging to a community that seemed to have remained intact across centuries. Overlooking the port, in the middle of a small square, I found a curious monument: a life-size statue of a mother monk seal with her pup. The figures were standing not far from where a local fisherman, back in 1975, gunned down the last two monk seals of a formerly healthy colony of the Egadi, laying the final tombstone over the species' presence in the western Mediterranean. The transition between a final act of destruction and the erection of a statue commemorating the former presence of such a charismatic mammal on the islands is a testimony of the momentous change in attitude towards monk seals across a handful of decades, not only amongst urbanites who may have idealised the conservation of marine species without perhaps ever having seen a seal, but also in the very places where people have to deal with nature on a daily basis. This new attitude is very much connected with the species' recovery we are currently witnessing in Greece, Türkiye, and Cyprus, where seals continue to be killed by some fishermen, but at a slower rate than in the past.

---

[17] WWF. Towards 2020 Report: https://tinyurl.com/3nc46mhx.

Are monk seals going to eventually reconquer their lost territory in the western Mediterranean? Contrary to what I might have answered this question even a few years ago, I am happy to affirm now that I am sure they will. Monk seals, as of recently, have been seen quite often in the Mediterranean outside their eastern strongholds, for instance, all around the Italian Peninsula, and not only in the Adriatic Sea but also in the Tyrrhenian Sea, including in Sardinia, and even further north in the Ligurian Sea.[18] And although these reports very likely concern lone wanderers and represent a far cry from the stable and productive colonies that still exist in the eastern Mediterranean, they provide nevertheless a very comforting message that the species has the potential for a comeback. The Egadi Islands could offer a strategic stepping stone for the monk seal reconquest of the western Mediterranean because they stand at the gates connecting the two parts of the basin and still contain suitable monk seal habitat within the boundaries of a protected space. Lone monk seals were recently seen in the area, the last of which was "captured" through a camera installed in one of the islands' marine caves in January 2016—no permanence, no certainty of breeding yet, but still presence: a good start.

Day 33: Marettimo, Italy

Marettimo is not a place to touch upon and then go without having, at a minimum, spent some time accepting the island's invitation to visit and explore. There is something magic around Marettimo that demands to be experienced, and I was not going to miss the opportunity, albeit a bit too fleetingly.

Leaving port early in the morning, having replaced a still-warm Moka in the galley closet, I sailed north along the coast for just over a mile to reach the Rock of the Camel, just in front of the cave with the same name (I have no idea about the connection with camels). This is one of the many marine caves in Marettimo but probably the most spectacular one, with its vast opening and the inside chambers, sandy beaches, and pools shimmering with the azure reflections of the water. It was impossible not to imagine the cave as it must have been a century ago, populated by families of monk seals.

The next stop was a few hundred metres beyond the Camel complex, under Marettimo's northeast headland called Punta Troia. Having left *Pontoporia* at

---

[18] Mo G. 2011. Mediterranean monk seal (*Monachus monachus*) sightings in Italy (1998–2010) and implications for conservation. Aquatic Mammals 37(3):236–240. https://doi.org/10.1578/AM.37.3.2011.236.

anchor in the small bay under the promontory, I hiked along the short path that leads to the summit, dominated by an ancient castle now recently renovated, where I could enjoy a breathtaking view of the island's eastern and northern cliffs. In the old Kingdom of the Two Sicilies, the castle was a prison. Below it, an infamous well was carved in the rock where the prisoners were lowered and kept in dreadful conditions. The castle has now been declared by some dreamers the seat of a "Monk Seal Observatory", a wishful idea given that there are not still many seals to be observed in the waters below—but charming nevertheless. What was that tiny disturbance of the sea surface I just caught with the corner of my eye, there in the distance? A monk seal, perhaps? Maybe not, but who could tell? These critters are always so elusive and enjoy hiding under the water surface.

Back aboard *Pontoporia,* I first doubled Punta Troia and then again Punta Mugnone, the northwest tip of the island, to sail for some distance counterclockwise along the coastline under the striking, steep cliffs of red dolomite plunging in the bluest and most transparent of marine waters. Thanks to the particular geology of this part of the island, wind, rain, and waves have carved the rock face across aeons into a series of the most astonishing caves—Grotta Perciata, Grotta Presepe, Grotta della Bombarda—which looked more like the entrances to cathedrals than mere holes in a cliff—natural cathedrals once populated by their rightful phocine clerics, as they would still have if it were not for the hostility of their human neighbours. I could not think of a more perfect monk seal habitat waiting to be inhabited again, and I wished them to be a beachhead for their reconquest of the entire western Mediterranean.

It was time to leave once more. Time to bid goodbye to the Strait of Sicily and to point *Pontoporia*'s stem towards the open sea and to Sardinia. This involved crossing the Sardinia Channel, almost 200 miles wide, requiring about 30 h of navigation.

The weather was fine, and the distance from the coast during the crossing kept me away most of the time from the breeze generated by land, so I had to turn on the engine again.

I had learned that the area I was sailing across had been earmarked for being the site of a vast offshore wind farm, likely to be installed here in the not-distant future. This being an area too deep for the installation of the more conventional type of turbines, involving towers sitting on the sea bottom, the wind farm here would be constructed over floating structures anchored to the bottom several hundred metres below. Whilst initially cringing at the idea of a pristine swath of sea surface becoming saddled with dozens of towers higher than the Tour Eiffel churning air with their gigantic blades, I am urging myself to think again—knowing that by doing so I might cause more than one eyebrow to raise amongst my more ecologically intransigent readers. In want of robust scientific proof to the contrary, I have been convincing myself that a floating wind farm is the least of a long list of evils within the range of contrivances humans have been devising to produce the energy they are—we all are—so hungry for. First, the energy produced is clean because it is carbon-free, so that it will not contribute to global warming. Second, unlike fixed wind farms on land or in marine shallows, a floating wind farm is removable at will, and once removed, it leaves little sign of it ever having existed. Third, it can be placed far enough from any coast to avoid providing visual pollution for people on land. Fourth, fishing will not be possible within a floating wind farm, which in this way becomes de facto a marine reserve, in addition to providing supplementary habitat for all the pelagic fish species that like to aggregate under floating structures. With noise produced by the turbines predictably below noxious thresholds, the farm is unlikely to exclude cetaceans from the area, particularly if it becomes an oasis against fishing. Granted, the farm must be carefully located away from significant flyways to avoid placing the deadly turbine blades in the way of migratory birds, and obviously, it must also be placed far enough from the main ship lanes, not to become a navigational hazard.

There was not much wildlife to run into on that day, except for an encounter with a small school of striped dolphins that appeared from the port side all of a sudden, out of nowhere, and remained with me for a dozen minutes, having great fun with *Pontoporia*'s bow wave. Common as striped dolphins might be in the Mediterranean, meeting them is always a joy, and their elegant beauty is always a gift to be thankful for.

Day 34: Sardinia Channel, Italy

In the early afternoon of the next day, I was hugging the southern coast of Sardinia, and by doubling Cape Teulada, the island's southwestern corner, I was able to point *Pontoporia*'s bow to the north, along the coast of the island of Sant'Antioco, until the island of San Pietro came into view. With the pleasant sea breeze that had come up around noon, it did not take more than a couple of hours from that point to reach the port of Carloforte, San Pietro's chief town, and my destination for the day. It was still well ahead of the peak tourist season, and finding a dock to tie the boat to was not difficult.

8

Pillars of Hercules: From Carloforte to Tarifa

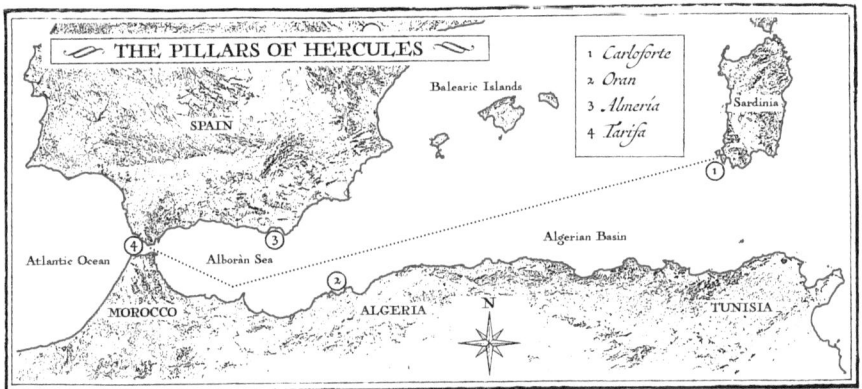

Illustration by Massimo Demma

Day 35: Carloforte, Italy

I woke up inside the well-protected harbour of Carloforte after a night of deep, dreamless sleep that healed me from all the nights and days I had been forced to stay awake and sleep piecemeal. It felt good to be back in the attractive small town of Carloforte, which has quite an intriguing history. Back in the sixteenth century, the world was still empty enough so that any group of people could decide to move to some place of their liking, accommodating the human natural habit of migrating without upsetting too many eventual neighbours as is the case today. In 1540, a colony of Genoese folks hailing

© The Author(s), under exclusive license to Springer Nature Switzerland AG 2024 **175**
G. Notarbartolo di Sciara, *Sailing Across a Wounded Sea*,
https://doi.org/10.1007/978-3-031-54597-9_8

from Pegli, one of Genoa's neighbourhoods to the west of the Ligurian city, decided to move to the island of Tabarka, off western Tunisia, having been invited there by the Bey of Tunis. The habitants of Pegli were excellent divers and were particularly good at collecting the treasured red coral, and the surroundings of Tabarka still had lots of the prized stuff. There is nothing particularly novel here, with countless small enclaves scattered across the Mediterranean coasts where communities from afar had implanted themselves seeking new opportunities or running from some disaster at home—an invasion, persecution, or an earthquake. Most of these enclaves are still in the same place to this day, in many cases recognisable from their original language retained through the centuries, hardly contaminated by their surroundings. The Genoese seemed to be doing fine in Tabarka. So they settled there to stay, except that 200 years later, they decided it was time to move on as their relationship with the locals had gone sour. In the process they had managed to clean up the area from the coral—a perfect example of unsustainable exploitation of a marine product well before being "unsustainable" had become the fashion of more recent times. In 1738, many of these Genoese moved to the Island of San Pietro in Sardinia, encouraged by Carlo Emanuele III, King of Savoy, who ruled on that island and was keen on settling a portion of his underpopulated kingdom. As it turned out, running from Tabarka was an excellent idea because soon after that, the Bey of Tunis decided to invade the island and enslaved the Genoese who had remained behind and who had to be ransomed by various European Christian rulers, including, again, the king of Savoy. Reunited in Carloforte, appropriately named in honour of their patron and saviour, the Tabarka refugees resettled so well in their new home that they are still there to this day—and speaking with a Genoese accent on top of that, which had proven as stable across centuries as their resolve not to return to Pegli. And why would they, given how pleasant Carloforte is? I would have been quite glad to spend a few more days, but soon, I had to resume my journey, for I still had a lot of ground to cover.

Before leaving, I wished to visit the tonnara of Carloforte, one of the few remaining in the Mediterranean. A tonnara—a traditional coastal tuna trap—is much more than the mere system of nets, elaborate as it is, that is placed in the water in mid-spring along many places of the Mediterranean coast, including here in Carloforte, to intercept and trap the mighty bluefin tunas as they migrate from the Atlantic into the Mediterranean for their breeding liturgy. A tonnara is also the site on land where the fish processing plant is built, and the boats and nets are stored off-season; it means work for tens of families centred on a millennial tradition based on the exploitation of the tuna: both the tradition and the great fish being today at risk of extinction.

The Carloforte tonnara is located at the northern tip of the island of San Pietro, in a locality called Isola Piana, about one hour's walk from the harbour. When I reached the location of the tonnara, all I could do was peep into the property through the closed gate since the fishing season had ended already, and the premises were empty of people and activities. But it was a pleasant walk amongst the low, scented Mediterranean scrub, with a fine view of the sea passage separating San Pietro from the main island of Sardinia: the same passage used by the tunas along their migrations, which brought so many of them here to their demise.

The Atlantic bluefin tuna is the biggest and probably the most famous representative of the tuna family, which consists of several species—15 in total, but 8, if we only count the larger ones, comprised within the genus *Thunnus*.[1] Whilst all the world's oceans host one species or the other, in the Mediterranean, it is the Atlantic bluefin (another silly name, as its fins do not seem to have anything particularly blue about them) that is the protagonist here. A magnificent, powerful fish, "the king of all fish" as defined by Ernest Hemingway, the most distinguished of all the fish butchers; a fish that spends its life cruising along its oceanic migration routes at high speed, in large schools like packs of 2-m long warm-blooded torpedoes, but known to exceed at times 4 m and a weight of almost 700 kg. The main nemesis for herrings, mackerels, and sardines.

Humans in the Mediterranean have learned to take advantage of these fishes since the dawn of times, exploiting their migratory behaviour to trick them into entering labyrinth-like netted traps placed on their paths and catching them in large numbers. Throughout the time, the Phoenicians, and then the Moors, and then various other Mediterranean peoples who inherited their techniques, such as the Spanish, the Sicilians, and the Sardinians, engaged in harvesting the bounty offered to them by such a generous sea, year after year and millennium after millennium.

Bluefin tunas are the quintessence of a pelagic fish, and in fact, they remain in the open sea for most of the year. But when they come into the Mediterranean from the Atlantic to breed, they change habits, and many of them reach their breeding grounds by swimming along the coasts in many locations. By doing this, they have become available to be captured by the coastal peoples for millennia. Special nets had been devised, forming a series of chambers and anchored off the coast in places such as Trapani, Favignana, Capo Passero, Formica, Bonagia, Scopello, San Vito lo Capo, Capo Granitola only to name the main ones in Sicily, but also in Sardinia such as here in Carloforte, and

---

[1] About the Tuna family: https://tinyurl.com/yf753w9f.

further to the north of the island, in Stintino. And, of course, also along the coast of Spain, Portugal, Morocco, and even Croatia. Swimming along the coast, unwittingly, the tunas enter the first chamber, and from there into a second one, then a third, and so on, where they are kept until the tonnara chief—i.e. the *rais*, Arabic word for "chief"—decides it is time to close the trap because it contains enough tuna, and the "mattanza" begins. The tuna is then driven into the last chamber, aptly called the "chamber of death", where the unfortunate fishes are brought to the surface and hauled up with special gaffs whilst the water becomes red with blood.

Called tonnara in Italy and almadrabas in Spain, traditional coastal tuna traps have become an important human activity with significant nutritional, economic, social, and cultural ramifications locally. Still, they never managed to dent the numbers of the tunas as they were entering the Mediterranean from the North Atlantic during spring to lay and fertilise their eggs. Those were the times of plenty and were supposed to remain such forever, until a handful of decades ago humans decided they wanted to catch more tunas.

Mediterranean tuna fishing started to go awry towards the end of the twentieth century due to various factors, including the enormous worldwide increase in the popularity of tuna meat, particularly when fed to the Asian masses in the form of sushi or sashimi. Increased exploitation was triggered by new methods to catch tunas anywhere they occurred in the open sea, and not just when they were hugging the coast as they migrated.[2] Financed by huge economic margins secured by these developments, highly technological vessels started ploughing the seas, tracked the tuna sometimes even with the help of spotter planes and sonar, and scooped whole schools out of the water in one bang with state-of-the-art purse seiners. Catches included not just large, mature individuals, but fishes of all sizes. These could be kept in floating pens for months and fattened with wild-caught feed before slaughter. The great majority of the tuna caught in the Mediterranean, up to 80% according to some sources, is being deep-frozen and sent by plane to the fish market in Tokyo, where a single 278-kg tuna recently fetched the eye-watering price of over three million dollars. With bullion swimming around in the sea available to anyone able to join the feast, the effects of the craze did not take long to become evident. By the mid-1990s, the Atlantic bluefin tuna was reduced by 85%. Astonishingly, under pressure from the industry and politicians, the regulatory organ in charge kept setting catch limits several times higher than those recommended by scientists, and even those were not respected. Bluefin tuna was being eaten to extinction.

---

[2] About the Atlantic bluefin tuna: https://tinyurl.com/5f5vrnah.

In the end, the Mediterranean population did not crash. Greed and mis-management had caused significant public outrage, and common sense eventually prevailed. Catch limits were set more realistically, and compliance was improved.[3] As a result, the Mediterranean Sea's incredibly generous, forgiving, and—in my opinion—undeserved nature is now allowing an increase in catch levels, hopefully sufficiently contained to ensure that the activity remains sustainable. Inside all this turmoil, one of the casualties has been the tonnara. The numbers of tuna migrating along the coastlines as they used to do in the aeons before the present, and available to become the quarries of the traditional coastal traps, had become too small to make most such activities profitable. The last mattanza in Favignana occurred in 2007. In Carloforte, the tonnara has survived to this day, undoubtedly helped by the slight uptick in the tunas' population size caused by more sensible management of the overall fishery during the last decade.

Back to town, it was time to get ready for the next passage, another long one—more than 700 nautical miles—which would bring me in about four full days to the western end of the Mediterranean Sea, to the Strait of Gibraltar and the gate to the Atlantic. A brief foray to the many shops from the streets lining Carloforte's harbour resulted in *Pontoporia*'s galley becoming enriched with delicious small Sardinian lentils, a couple of whole Sulcis pecorino cheese, a bottle of local juniper grappa (a very aromatic Mediterranean version of gin), and a jar of pesto, a clear legacy from the Genoese origin of the Carlofortini.

Meanwhile, after a few days of high atmospheric pressure, the onboard barometer had taken a dip, and a warm front had arrived over southern Europe with a tail extending to the central Mediterranean, generating a moderate Sirocco wind, an overcast sky, and—for the first time since the Neverin in Rovenska—even a few drops of rain. The swell generated by the wind would have significantly impaired my ability to see wildlife on my way towards

---

[3] The Atlantic bluefin tuna, *Thunnus thynnus*, is listed as "least Concern" in IUCN's Red List: https://tinyurl.com/54yu67yv.

the Strait. Still, the wind's south-easterly direction suited me just fine because it would have pushed *Pontoporia* forward without resorting to the engine's help.

As I exited Carloforte's jetties and turned south to hug San Pietro's coast for about 3 miles, I was still in the lee of the nearby island of Sant'Antioco. But after reaching Punta delle Colonne at the island's south end, I continued south for a mile to steer clear to the shoals and admire the dramatic rocky pillars (the "Colonne") towering above a swelled sea that had turned from blue to grey under the clouded sky.

Moments later, the way ahead clear of dangers, I set the course to west-south-west, and *Pontoporia* forged forward on a free reach, the wind singing in the halyards, heading for Tarifa, Spain. I was on my own again, and I had the impression of sharing a joyful sense of freedom with the accompanying European storm petrels[4] that had replaced Scopoli's shearwaters as my bird companions on that day, and were fluttering around *Pontoporia*.

Days 36, 37: Algerian Basin

After more than a month of sailing, the sea had morphed into something very close to my natural habitat. It was not just *Pontoporia*'s microcosm; it was the sea itself, which is not so natural given that humans are land animals, but such is the power of adaptation of the species I belong to. Queasiness caused by motion sickness—a pretty uncomfortable sensation that usually rears its ugly head after long periods spent on land—was long forgotten. My body was anticipating the movement of the waves without me noticing, and I was moving around the boat like in a dance, gravity being no longer the only force to contend with. It was as if body and mind had agreed to adapt to the different lifestyle, with the frequent alternance of rest and vigil becoming the physiological norm. The sense of being safe in the elements was absolute, with only the perspective of being run over by a ship occasionally putting that sense to test.

The sea had provided me with bad experiences before. Like getting trapped in storms far from shore in situations that had successfully tried to scare me with the violence of the waves, the threatening darkness of the skies, and the roaring noise of air rushing over water. I had even lost a boat in a fire in the 1980s when sailing solo in Mexico, a circumstance that forced me to regain the safety of shore rowing on a small dinghy for hours during the dark. All of that generated a healthy sense of respect for the ocean and awareness of my

---

[4] The European storm petrel, *Hydrobates pelagicus*, is listed as Least Concern: https://tinyurl.com/4dmukep4.

frailty against elements that are infinitely stronger than me. But in the process, I learned the tricks of survival, to a large extent, by paying attention to the signs received from my physical surroundings, and behaving accordingly by avoiding recklessness. Technology had helped significantly by giving me constant and exact information about my position concerning the dangers of land and the whereabouts of other vessels, and by providing increasingly reliable weather forecasts that had made it almost impossible to get caught by surprise by the deceptive elements.

That day, I knew things were all right as *Pontoporia* smoothly rode from crest to crest under the cloudy sky. Twenty-four hours had elapsed since I left Carloforte, and under the push of Sirocco, I had already gone 210 miles, which was a 25% gain over my anticipations. I was as far from land as I could be in this part of the western Mediterranean, with Menorca and the Balearic Islands almost a 100 miles away to starboard and the Algerian coast about 60 miles off to port. The sense of commonality with the marine birds that were a constant company was of the physical type, animal with animal, far from being a pseudo-literary attempt to force poetry into my narration to make it sound more intriguing.

I prayed to the god Eolus to keep the Sirocco blowing as long as possible, which was making *Pontoporia* go faster, thereby significantly shortening my travel time. This help was more significant, considering that as I was getting closer to the Strait of Gibraltar, I would have had to sail against a current that would have slowed my travel. Like tides, surface currents are rarely a navigational issue in the Mediterranean, except when approaching passages constrained between lands. For example, crossing the Strait of Messina could be so challenging that the ancients believed it to be guarded by the sea monsters Scilla and Cariddi. The Strait of Gibraltar connects two bodies of water much larger than the Strait of Messina, but it is much broader and poses different challenges to the sailor who crosses it.

It should be no surprise that oceanic water masses are in constant motion. When sailing over the ocean, we are doing something quite different from moving on land. We should think of the ocean not simply as an inanimate mass of water but as a breathing giant capable of moving from one side to the other of the planet and determining the fate of countless beings—including fishes, birds, mammals, turtles, and the multifarious components of the plankton—that have been interweaving for aeons their lives and movements to the rhythm of such breath. Two are the main physical forces that drive water masses around: the friction of wind over the sea surface that pushes the waters in its direction; and the dynamic effects of different water densities derived from heat exchanges with the atmosphere.

Ocean circulation is a complex subject that I wish to leave to physical oceanographers to explain in detail, and one that would have little traction in these pages. But the Mediterranean is a particular case with an interesting story to be told. It is a story that starts with questioning why one has always to fight an opposing current, sometimes more than three knots in velocity, when sailing from the Mediterranean to the Atlantic across the Strait of Gibraltar—as was my case during this trip. This incoming current is caused by water sucked into the Mediterranean from the Atlantic to compensate for the amount that evaporation constantly subtracts from the basin: water that gets from the sea into the atmosphere as vapour and is then carried away. The Mediterranean lies under a mostly dry atmosphere with a very high evaporation rate; only a third of the evaporating mass is replaced by freshwater falling back in as rain or flowing into it through the few relatively unimpressive rivers running across the continents surrounding the region. The Nile is not at all unimpressive, as we have seen, so it is an exception to the rule, albeit not as much as it could have, had it not been thwarted by the Aswan High Dam. Some water flows into the Mediterranean from the Black Sea through the Dardanelles, but not much. Ultimately, the resulting negative budget can only be compensated by the inflow of Atlantic water through Gibraltar.

But the situation is more complex than that. As Atlantic water flows into the Mediterranean along the surface, forming a thin layer—no more than a few hundred metres thick—and progresses eastward, some of it escapes to the atmosphere through evaporation, as we have seen, thus becoming saltier in the process. By the time this surface water reaches the dead end of the Levantine Sea, it has become denser and heavier, ending up quite different from the original. Pushed back by land, by being heavier it slides under itself without mixing with the incoming flow, just like oil will not mix with water. This countercurrent, called "Levantine Intermediate Water", is about 300 m thick and creeps back into the Atlantic over the shallow sill of the Gibraltar Strait.

To complete the story, I should mention a third layer of water, the densest of all: the "Mediterranean Deep Water". Formed during the coldest days of winter by the combined effects of cooling and evaporation in the northernmost parts of the basin, such as in the north Adriatic and north Aegean seas, this water sinks and slides onto the very bottom of the basin, where it stays because it cannot rise high enough to trickle out of the Gibraltar Strait.[5] The scheme just described is an oversimplification. For example, as the Atlantic water rushes into the Mediterranean, it doesn't head straight to the Levantine Sea. It breaks into gyres, eddies, and meanders along the way, a testimony to

---

[5] About Mediterranean hydrologic features and climate: https://tinyurl.com/33uwh675.

the fundamental interference of chaos in all things natural. This element of disturbance has significant consequences for the ecology of the Mediterranean and of the Alborán Sea in particular.

As the second night approached, I had travelled more than 400 miles from Carloforte. Where I was, the coasts of Europe and Africa were converging, now less than 100 miles apart, with Cabo de Palos in Spain to starboard and the Algerian coast leading to Oran to port. The Sirocco was still up, but it showed some signs of subsiding. Meanwhile, the barometer was also rising, and there was a break in the cloud cover to the west, right where the sun was getting ready to set in a kaleidoscope of colours, whilst a first-quarter moon was peeping in and out from behind the scattered clouds.

Pontoporia's pace had slowed, but she still kept a respectable average of seven knots under sail. As the wind was losing impetus, the prospect of continuing under engine was becoming more and more likely. Still, I was comforted by the thought that this same current that was against me now would have turned in my favour on my way back, as I approached the conclusion of my journey.

Day 38: Alborán Sea

Before the dawn of the third day of navigation from Sardinia, I had entered the Alborán Sea, the Mediterranean antechamber to the Atlantic Ocean. The Alborán Sea is unique because—as the conduit between two worlds and the intermediary of titanic exchanges of water masses—it is the most dynamic portion of the Mediterranean. Strictly speaking, from an ecological point of view, the Alborán Sea waters are more similar to those of the Atlantic than those of the Mediterranean. In other words, whereas the geographic boundary between the two is situated at the Strait of Gibraltar, the ecological boundary, somewhat counterintuitively, is placed much further to the east, well inside the Mediterranean proper. As the Atlantic water rushes in across the Strait, it creates a clockwise gyre between the coasts of Spain and Morocco, and as soon as the water gets out of that first gyre, it makes a second, identical one just to the east of it. As the water in the second gyre flows from north to south, it creates a permanent front separating the Atlantic from the Mediterranean water masses along an ideal line connecting the cities of Almeria in Spain and Oran in Algeria. The ecological boundary between the Atlantic Ocean and the Mediterranean Sea is created along this oceanographic front.[6]

Being highly dynamic for a marine ecosystem is almost invariably a synonym for being highly productive, because of the greater mixing that can occur between deeper, nutrient-rich waters with the surface. This mixing is conducive to triggering life processes at sea, and high productivity is, in fact, a characteristic of the Alborán Sea. This condition, in turn, supports the subsistence of large and diverse populations of marine organisms—including cetaceans.

Such richness was amply documented by colleagues Ana Cañadas and Ricardo Sagarminaga through data they painstakingly collected for years aboard *Toftevaag*, a picturesque 1910 Norwegian fishing boat that Ricardo had converted into a research vessel.[7] I remember briefly boarding the *Toftevaag* when she was docked in the harbour of Valencia during a conference I attended in the Spanish city, imagining the hundreds of days that Ana and Ricardo had spent on the venerable craft and the extraordinary images of multitudes of cetaceans appearing in front of their eyes in that part of the Mediterranean, one of the most generous in terms of such encounters. Since the early years of their activities, they have reported 270 sightings of striped dolphins, 254 of common dolphins, 171 of pilot whales, and 125 of bottlenose dolphins, all of them collected by zigzagging within the Alborán Sea over a distance equivalent to four times the length of the Mediterranean. As a result of their research effort, they were able to characterise with robust scientific data the habitat preferences of the different species, paving the grounds for establishing MPAs in the region.

In subsequent years, and largely thanks to their effort, three IMMAs were identified in the Alborán Sea,[8] testifying to the area's importance for Mediterranean cetaceans.

---

[6] Tintore J., La Violette P.E., Blade J., Cruzado A. 1988. A study of intense density front in the Eastern Alborán Sea: the Almeria - Oran Front. Journal of Physical Oceanography 18:1384–1397. https://tinyurl.com/575pfkxj.

[7] Cañadas A., Sagarminaga S., García-Tiscar S. 2002. Cetacean distribution related with depth and slope in the Mediterranean waters off southern Spain. Deep-Sea Research I 49:2053–2073. https://doi.org/10.1016/S0967-0637(02)00123-1.

[8] The "Alborán Sea IMMA" for bottlenose and common dolphins (https://tinyurl.com/527wzj6y); the "Alborán deep IMMA" for pilot whales, sperm whales, Cuvier's beaked whales and Risso's dolphins (https://tinyurl.com/2s3fjdd6); and the "Alborán corridor IMMA" for sperm and fin whales (https://tinyurl.com/2dccnpja).

By the time the sun had come up, I was already well into the Alborán Sea, and the Sirocco wind had, in the meantime, ultimately died out so that I had to wake up *Pontoporia*'s auxiliary engine to proceed; but not for long. A good thing about the presence of the gyres in the Alborán Sea was that by diverting my course to navigate closer to the African coast, the current was now in my favour. This condition, added to the benefit of the sea breeze that had come up around noon due to the vicinity of the Moroccan landmass, allowed me to proceed quite well under sail and generously compensated me for the longer distance the change of course required.

So here I was sailing in the famously rich Alborán Sea, particularly the "Alborán Sea IMMA" identified in recognition of its importance for both bottlenose and common dolphins.[9] I was scanning the horizon left and right, hoping to benefit from such a bonanza by enjoying some memorable encounters. Sighting conditions were not great because the breeze created by the differential temperature between the cool sea and the hot African continent had caused the sea surface to be moderately choppy: not what I would have defined as perfect conditions to detect the presence of animals at the surface. Yet, what I was about to witness would have been evident in the meanest of storms.

I knew I was running into something big from the first, distant telltale splashes on the horizon. The perception of the presence of animals in the distance has much to do with the "normal" behaviour of water. Although I am fully cognisant of the inappropriateness of attributing behaviours to an inanimate medium, I have no better way of expressing myself. The eye acquires images of the constantly moving sea surface and allows the mind to create a catalogue of the many different ways these motions occur as a result of the dynamics of the inanimate water masses: waves, wave crests, different wave trains crossings, spindrifts, splashes created by breakers, and the infinite variations of the theme that, however, have one thing in common: they are all caused by the inanimate forces of water of air, and of the two interacting. After a whilst, one can understand, without even thinking about it, that this range of behaviour is "normal" and, therefore, doesn't require any particular

---

[9] Alborán Sea IMMA: https://tinyurl.com/527wzj6y.

level of attention. But place a small critter somewhere in the distance and have it make even a tiny splash with its motion visible from above the surface. If your eye catches it, the alert is given: there is something alive out there, something more exciting than just water. And if there is one, good chances are that there is more.

On that day, there was something quite alive in the distance, and there was a lot of it. As I was getting closer to the commotion area, the cause of such commotion became more evident because, besides the splashes, bodies of marine animals had become visible, some even briefly jumping out of the surface. Approaching the spot, bits of information were trickling in: whatever these critters were, they were feeding on a shoal of small fishes, as indicated by marine birds circling, fluttering excitedly, and diving above them. And the movements of the predators at the surface were the movements of dolphins, not tunas: the latter only making fleeting appearances, with the tips of their dorsal fins cutting the surface like swords as the bigger fish race in pursuit of their prey, but rarely giving the observer the satisfaction of a good view of the whole animal. Not so the dolphins, which at that point were well identifiable with their falcate fins going to and fro and frequently jumping clear of the water in acrobatic somersaults.

As I came within a distance of about a 100 m from the action, all details had become clear. This was a huge school of common dolphins, maybe two or three hundred of them. When the animals are so many and moving in all directions around you, most of them being underwater with only a fraction of the group at the surface, it becomes challenging to make even a rough estimate of the size of the school. Hundreds of common dolphins in a single group, once upon a time perhaps a not unusual occurrence in the Mediterranean, are today something unheard of in the rest of the region, attesting to the persisting uniqueness of the Alborán Sea and connected with the availability here of abundant prey.

Common dolphins in the Alborán also had different habits from their conspecifics further inside the Mediterranean, east of the Almeria-Oran marine boundary. Here, these dolphins prefer deep, offshore waters like their oceanic relatives, in stark contrast with the inner Mediterranean common dolphins that are predominantly coastal creatures, as we have seen when encountering them in the relative shallows of the Ionian, Aegean, and Levantine seas. Their different ecological habits in the east are also reflected in genetic differences that became evident when comparing dolphins from these different geographic areas, a telltale of the fact that the inner Mediterranean population has been isolated long enough and that there isn't a great deal of genetic exchange amongst common dolphin "tribes" across the basin.[10]

As I turned on the hydrophone, I could hear the noisiest possible kerfuffle of whistles, clicks, creaks, and click trains that the dolphins were emitting as they communicated with their schoolmates whilst racing through the sardine shoal in deadly dashes. With time, as the dolphins' stomachs were getting fuller and the shoal of sardines during the process probably too small to remain interesting, the animals became calmer and swam around more leisurely. All the birds had gone somewhere else to find more stuff to lay their beaks on, and it was time for the dolphins to relax. A small group of them took an interest in the presence of *Pontoporia* nearby and came to inspect, as dolphins often do. Coming at close quarters, I could have a better look at their elegantly decorated flanks, and—surprise—I discovered that there were quite a few striped dolphins in the mix. Such mixed groups, as I found out later relating my experiences to colleague Ana Cañadas, are far from unusual in the Alborán Sea. Remembering the stories about striped dolphins in the Gulf of Corinth in Greece, encountered earlier during my journey, I noted that there must be some sense of affinity bringing together the two species. In the Gulf of Corinth, striped and common dolphins not only team together but also mate and produce fertile offspring.

The dark was descending, and it was time to resume my course towards the Strait of Gibraltar. Maritime traffic was getting intense again as I was, once more, heading northwest and approaching the main ship lane crossing the Strait, with the coasts of the two continents bordering the Mediterranean converging towards the narrows. However, the weather was fine, and all the instruments aboard provided sterling service.

Day 39: Alborán Sea

At dawn, I was no more than 50 miles from my destination, and land was in sight all around. The day was clear, the visibility high, and distant lands were in view. To the south, no more than 50 miles away, Morocco's tall Rif mountains stood out with a shade of blue just darker than the sky above; to the north, the hills of Andalusia were rising inland from Malaga. Soon, the

---

[10] Natoli A., Cañadas A., Vaquero C., Politi E., Fernandez-Navarro P., Hoelzel A.R. 2008. Conservation genetics of the short-beaked common dolphin (*Delphinus delphis*) in the Mediterranean Sea and in the eastern North Atlantic Ocean. Conservation Genetics 9:1479–1487. https://doi.org/10.1007/s10592-007-9481-1.

converging coasts of Africa and Europe would have appeared to the west ahead as I approached the Strait of Gibraltar.

The Strait is more than just a point of the Globe and the quadruple meeting point between two ocean basins and two land masses. As far as mariners are concerned, the Strait is a historically relevant point of passage between two worlds. In the old days, the passage was wrapped up in a thick mythological apparatus for us Mediterraneans, nothing less than between the known and the unknown. What was beyond the Strait was anyone's guess, but it also was something strongly intimidating that sane people would do well to stay clear of. Ulysses was said to have ventured into the Strait's surroundings as part of his farthest peregrinations, and the island of Ogygia, the golden cage where the nymph Calypso had kept him for 7 years, was there somewhere: well into the Atlantic according to Plutarch, or still in the Mediterranean according to others, but at the boundaries of the known world anyway. Plato suggested that strange worlds, such as the fabulous island of Atlantis, could be found somewhere beyond those frightful gates.

The main image, inherited from those old tales, is one of a Strait guarded on both sides by two towering structures, the Pillars of Hercules. A stern warning was purportedly embodied in the Pillars: "ne plus ultra", go no further. Not that the warning seemed to prevent the fearless Phoenicians from exploring and trading beyond, as they fancied. The pillars were said to meet the Mediterranean sailor rising on either side of the 10-mile-wide Strait: Calpe Mons to the north, today known as the Rock of Gibraltar, and perhaps (historians are divided) Abila Mons, or maybe Jebel Musa, on the African side. Hercules had ample opportunities to exhibit his bravery when he was requested to reach the Garden of the Hesperides to steal some cattle from the giant Geryon and return them to a Mycenaean king. In his time, the legend was that the Mediterranean was separated from the Atlantic Ocean by the mass of the Atlas Mountains. Instead of reaching the Hesperides by crossing the mountains, Hercules chose instead to smash through the Atlas, creating a connection between the Atlantic Ocean and the Mediterranean, and in the process, the two parts of the split mountain remained in the form of pillars on both sides of the opening. I resist the temptation of regarding the stories of Hercules' feats in this region as a mythological recapitulation of the geological events that occurred there 6 million years earlier, when the Mediterranean had had its communication with the Atlantic blocked 6 million years before and then reopened by the movements of the continental plates. A charming idea to connect myth with geological reality, but one that I view as devoid of substance.

What is certain is that the mere idea of the Pillars standing guard over the gate of the Mediterranean inspired awe and respect in the ancient sailors. As far as I was concerned, knowing that what was beyond the Strait was not particularly more wondrous than what was within—and undoubtedly no more daunting—my excitement was of a different nature. With space telescopes orbiting around Earth and placing humanity's eyes in front of spaces and times that were unimaginable before, humans have managed in current times to push their world's boundaries incredibly farther, even beyond the Solar System. However, I think that the real boundaries that count today differ from those of a spatial nature. The main question has become whether humans will be able to engage in a relationship with their world that is respectful and sustainable, as opposed to the ability to conquer new worlds.

The source of my excitement as I was approaching the gate to the outer ocean had more to do with the realisation that, as far as the animals I was searching for were concerned, I was navigating within a boundary between two very different natural marine systems: the Atlantic where once everything came from, and the Mediterranean as the terrain for their conquest and colonisation, when waters from the ocean rushed back into the inland sea after the connection between the two had opened again by the movements of the Earth's crust 6 millions years ago. At that time, the animal species that came into the newly formed sea became regular residents there and later had sufficient time to diverge from their Atlantic forebears, having developed biological differences recognisable in their current genetic makeup.

These new invaders included, amongst many others, all the cetacean species currently found in the Mediterranean—fin and sperm whales, long-finned pilot and Cuvier's beaked whales, as well as common, Risso's, bottlenose, striped and rough-toothed dolphins—as well as the spinetail devil ray and the monk seal. Should we believe another version of the Hercules myth—that the Greek hero had erected his pillars to narrow down an existing passage to prevent sea monsters from the Atlantic Ocean from entering the Mediterranean—we can note with relief not only that Hercules failed in his purpose, but also and most importantly that the sea monsters of yore—the whales, the dolphins, the seals, the sharks, and the rays—are instead creatures that we love, we are concerned for, and that we wish them to flourish in the Mediterranean.

Not only Hercules' Pillars didn't manage to prevent the colonisation of the Mediterranean by marine fauna that succeeded in becoming established inhabitants of the region. The funnel-shaped morphology of the Spanish and Moroccan Atlantic coasts abutting the Strait, combined with a constantly inflowing surface current, acts like a gigantic lobster trap luring into the

Mediterranean a variety of foreign species. There is a rich collection of episodes in which complete strangers came into the Mediterranean, found on occasion as temporary voyagers who did not have the opportunity to become established in this sea with a viable population.[11] The ebullient humpback whales, for instance, once on the way to extinction under the punishing impact of twentieth-century commercial whaling but today fully recovered almost everywhere in the world, have now been appearing in the Mediterranean often, at the rate of more than once per year, and might one day decide to move in more permanently, were the conditions here to remain favourable to cetacean life. The late whale biologist Roger Payne, who in the 1960s brought the eerily beautiful song of the humpback whale to the knowledge of the broadest public with a disk made available in an issue of the National Geographic Magazine, imagined that the myth of the song of sirens described by Homer in the Odyssey might actually have been humpback whales present in the Mediterranean in antiquity, and long since disappeared for unknown reason.[12]

Much more unusual has been the appearance of grey whales inside the Mediterranean twice in the current century, which we had already noted when mentioning a famous sighting made off Israel a few years ago by Aviad Scheinin. But grey whales are a different story from humpback whales altogether because they belong to a species that, having disappeared from the North Atlantic in the early 1700s, is today found only in the North Pacific. Whilst keeping in mind that Earth is a small planet for these distant travellers, able to go a long way without much effort, the occurrence in the Mediterranean of grey whales remains a puzzle. It is not entirely unplausible that during their summer northward migration along the Pacific coast of North America, some of these whales might have ended up much farther in the north than usual due to the reduction of polar ice extent caused by global warming, thereby trespassing into more eastern waters north of the American continent, across the Northwest Passage in Arctic waters. Once the seasonal drive to migrate south again had kicked in, these whales might have found themselves in the Atlantic instead of in the Pacific, and by wandering southwards along the eastern shores, they might have ended up in the Mediterranean. Admittedly, this is a highly speculative explanation—but not an impossible one. Indeed, the Mediterranean, which facilitates movements and migration for a host of flying animals going along a north-south direction with its many land bridges

---

[11] Notarbartolo di Sciara G., Tonay A.M. 2021. Conserving whales, dolphins and porpoises in the Mediterranean Sea, Black Sea and adjacent areas: an ACCOBAMS status report 2021. Editions ACCOBAMS, Monaco. 160 p.

[12] Payne R. 1995. Among whales. MacMillan Publishing Company, 431 p.

and crossroads, for the animals that can only swim is basically a cul-de-sac, and ultimately a potential trap. A second grey whale has occurred in the Mediterranean in recent times. Judging from its size, this one was very young; its wanderings were followed with increasing concern by the media and the public. The little whale appeared, increasingly emaciated, along the coasts of Morocco, Algeria, Italy, France, and Spain but was no longer seen after a last appearance off Barcelona. We fear that it has starved to death.

As I approached the Strait with only a handful of miles to go, with the imposing rock of Gibraltar coming up at the starboard side, it was quite obvious where the pillar to the north side was meant to be. Although I had some trouble figuring out which of the various land heights beyond Tangier and Ceuta should be considered the southern pillar, it was not too much of an effort of imagination to see the two continents getting together at this point, emphasising with their presence in the powerful concept of the two masses guarding the gates to what was once considered the unknown world.

Eventually, as I had already crossed the narrowest point of the Strait and as the long-winded oceanic swell had begun to make its presence known by slowly lifting *Pontoporia* and then bringing her back down in a gentle undulation, the town of Tarifa came into view with the pretty Santa Catalina castle towering over it, and the low-lying islet of Tarifa, now no longer an islet as it had eventually been connected to the mainland with a causeway, guarding the entrance to the harbour. It was late afternoon when I docked in Tarifa's tiny, beautifully protected Puerto de la Rada. I had come to the southernmost tip of Europe, and at the same time the westernmost extreme of my journey.

9

Pelagos Sanctuary for Mediterranean Marine Mammals: From Tarifa to Sanremo

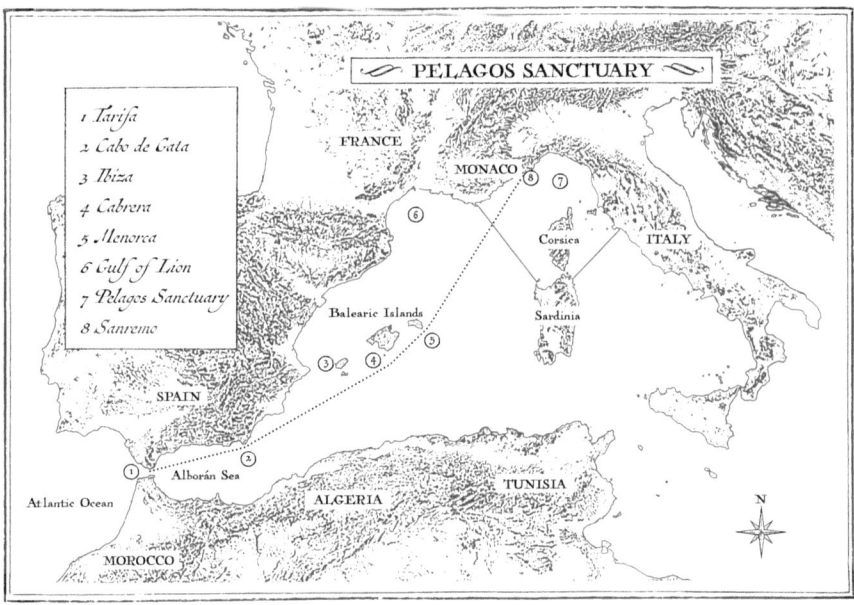

Illustration by Massimo Demma

Day 40: Tarifa, Spain

There was no dearth of reasons for my desire to sail all the way to the Strait of Gibraltar and venture into the Atlantic Ocean. The dense interface among myth, history, geography, oceanography, and ecology made the Strait supremely attractive. Still, one particular reason drew me to it like a magnet: the chance of encountering one of the most extraordinary marine animals, the orca.

The largest and mightiest member of the dolphin family, the orca is also known in English as "killer whale", a name I strongly dislike, and a rather silly one if I may say so. Another of these objectionable names originating from the days of whaling, derived from the acute observation that these cetaceans, lo and behold, kill their prey in the process of eating it. The orca's features are unmistakeable: a shiny black back with a tall dorsal fin, in striking contrast to a clearly delimited, bright white underbelly and a pair of conspicuous oval spots on the back of the head. Males are much larger than females and can reach 9 m in length. The dorsal fin of an adult male is toweringly vertical, up to 2 m in height, and looks like a war banner. The orcas' demeanour is one of awe-inspiring and extreme charisma.

Orcas have become very popular in recent years among the worldwide public because of their size, stunningly elegant colouration, and the image of power they convey. Beautiful, intelligent, and fearsome, they would have had no natural enemy had humans not existed. And yet, although they can best just about any other animal in the ocean, from the tiny herring to the white shark and even the colossal blue whale, they are remarkable for having never attacked humans, except for the rare incidents in the quite unnatural conditions created by orcas kept in captivity. Whether this peculiarity results from a mysterious tendency to extend their highly social behaviour to our species, we do not know. Whatever the reason, by abstaining from being predators of *Homo sapiens,* orcas are giving us a demonstration of wisdom. However, such wisdom has not exempted them from incurring merciless human persecution.

Orcas came recently to the attention of the broader public because of the controversy surrounding the plight of their kind in captivity. The spark was provided in February 2010 by the death of trainer Dawn Brancheau, killed by a male orca named Tilikum during a public show at Orlando Sea World in Florida. Tilikum had been in captivity since his capture near Iceland in 1983, and the news emerged that he was previously responsible for the death of two other trainers.

The story of Tilikum and other orcas in captivity was related to the 2013 movie, "Blackfish", which won acclaimed reviews and was broadly circulated.[1] Blackfish played an important role in influencing public opinion, leading many people to consider ethically unacceptable the entertainment industry's practice of keeping intelligent animals such as dolphins and orcas imprisoned for life inside tiny pools. An increasing share of the public came to realise that small concrete tanks cannot possibly be adequate for the well-being of captive orcas, not only in terms of the tanks' grossly insufficient size but also, perhaps as importantly, because of the deprivation by captivity of their most elemental social needs.

Orcas live in societies that are among the most stable of all cetaceans and perhaps of all mammals. Their pods typically comprise from several to about 20 individuals. Orcas usually remain lifelong with their natal pod; the females are the backbone of orca society, with grandmothers, mothers, and daughters connected within the matriline. Male orcas also remain in their natal pod except for brief forays to mate with females from other pods, when two or more groups occasionally coalesce.

Science currently recognises a single orca species worldwide, *Orcinus orca*. However, its many separate geographic populations display striking differences in size, colouration, and, most evidently, behaviour. Zoologists have yet to dare to dive conclusively into the complexities of orca taxonomy and propose the subdivision of the species into different subspecies or perhaps even species. This is likely to happen soon. For example, three "types" of orca are found along the Pacific coast of British Columbia in Canada, where the species has been studied intensely for decades. One of these types eats only salmon, and another eats only marine mammals. The third type remains further offshore and specialises in preying on sharks, and, in fact, the older orcas in that population often have their teeth worn down from habitually grasping their prey's rough skin. Salmon- and mammal-eaters often meet at sea (in technical terms, they occupy the same geographical range and are thus "sympatric"). Still, when their paths cross, neither seems to recognise the other's presence, as if the two tribes were invisible to each other.

Across the world's oceans, there are orca populations that are specialised in eating different types of fish (salmon, herrings, Patagonian toothfish, sharks, tuna etc.), and other orcas are specialist predators of mammals (seals, sea lions, various species of dolphins and whales, sea otters). For instance, the Strait of

---

[1] The documentary "Blackfish" telling the story of the captive orca "Tilikum" is described here: https://tinyurl.com/yudjvr7a.

Gibraltar orcas are bluefin tuna specialists; they know very well where and when to get them, as the mighty fishes punctually congregate at the narrows in spring and early summer to enter the Mediterranean during their breeding migration.

Orcas are found just about everywhere in the ocean, from polar to tropical latitudes, with a clear preference for the colder, more productive waters where the abundance of prey can sustain their hefty nutritional needs. They occur only rarely in the Mediterranean, although, ironically, it is from here that perhaps the oldest written account of an orca originates, dating to the middle of the first century CE. In his *Naturalis Historia*, Pliny the Elder tells the story of an orca who entered Rome's harbour of Ostia, supposedly attracted by hides that had fallen overboard. This reason seems rather unlikely to me, but it is an old story and interesting in itself. Regardless of the part concerning the hides, the fact that the animal entered the harbour is not so implausible, as demonstrated recently when a pod of orcas came into the port of Genoa.[2] Whatever the reason, the unfortunate cetacean was attacked by no other than the emperor Claudius, who slaughtered it with the assistance of his guard.

Orcas are easily identified at sea because of their conspicuous features. Therefore, the chances of detection are good when a pod enters the Mediterranean; so, we know today that the species rarely occurs here.[3] This is not the case, however, for the small ocean space between the European and African coasts in the Strait of Gibraltar, where they converge from their broader northeast Atlantic abodes at the beginning of summer. Their mission: feasting on bluefin tuna.

In this locality, orcas have been the subject of a long-term investigation by Spanish colleagues such as Ruth Esteban, who has catalogued them one by one based on the distinguishing marks on their bodies. Documented photographically, such markings have allowed the compilation of a catalogue of orcas forming a population consisting today of only 39 individuals. Ruth and her colleagues have also been able to record the frequent episodes of conflict with Moroccan hook-and-line fishers, whereby the orcas manage to wrestle hooked tunas out of the fishers' hands. Such episodes undoubtedly have not increased the orcas' popularity amongst the Maghrebi fishing communities, although, to be fair, the root cause of conflict is not the orcas who have been there for aeons, but the scarcity of tuna caused by overfishing perpetrated by the big industrial purse seine fleets in the Mediterranean. Not unlike the

---

[2] Rare Sighting of Killer Whales off Genoa Coastline: https://tinyurl.com/zpfcdsjk.
[3] Notarbartolo di Sciara G. 1987. Killer whale, *Orcinus orca*, in the Mediterranean Sea. Marine Mammal Science 3(4):356–360.

ruinous effect on indigenous peoples of the mindless destruction of bison herds in the North American plains, the equally mindless overexploitation of North Atlantic bluefin tuna has resulted not only in hardship for small-scale Moroccan fisherfolk but also in this population of orcas—assessed as Critically Endangered in IUCN's Red List possibly because of the depletion of their natural prey.[4]

I realised that the chance of finding such a small number of animals in the area was not too dissimilar from finding the proverbial needle in a haystack; nonetheless, I badly wanted to encounter these orcas, and it was the right time of year to find them. So it was that early that morning, I took *Pontoporia* for a day-long foray into the Atlantic.

After exiting the harbour of Tarifa, I sailed across the "Strait of Gibraltar and Gulf of Cadiz IMMA", identified in recognition of the area's importance for the orcas.[5] My plan was to hug the low-lying Atlantic coastal plain of Andalusia created by the Guadalquivir River estuary and head toward the city of Cadiz. I sailed along the coast until I was offshore from the town of Barbate, in view of the famous Cape Trafalgar, where the British fleet under the command of Horatio Nelson trounced the Napoleonic forces in 1805.

The day was overcast, with a grey sky above and a grey sea below, so different from the customary blue of the Mediterranean. Nevertheless, despite the rolling oceanic swell, the weather was calm, and the orcas would have been visible had they been around. Unfortunately, none were. Disappointed, I turned *Pontoporia*'s bow back towards the dock in Tarifa (Fig. 9.1).

Later that day, I contacted Ruth, who had moved to Madeira to study the rich cetacean fauna of that oceanic island, to ask whether she could give me more information about the orcas in the area.

"Why don't you get in touch with Alfredo Rodrigues?" she suggested, "Alfredo has very interesting stories about the local orcas".

---

[4] The Gibraltar population of orcas is currently assessed as Critically Endangered: https://tinyurl.com/4zzw7mu7.

[5] Strait of Gibraltar—Gulf of Cadiz IMMA: https://tinyurl.com/ms888bzk.

Fig. 9.1 A pod of orcas patrolling the waters off the coast of Algarve, Portugal. The animal in the rear of the group is an adult male, recognisable from its very tall dorsal fin. The drawing is based on images kindly provided by Alfredo Fernandes (Illustration by Massimo Demma)

Alfredo is a whale-watching operator based in Faro, a town in the Portuguese region of Algarve, near the border with Spain. Faro is located on the Gulf of Cadiz, where I had sailed that morning, and is well within the summer range of the orca population.

"Indeed, I do have stories to tell", said Alfredo, when I managed to reach him. "I've worked with a whale-watching company based in Faro since 2017, so I've had the chance to run into these magnificent animals during their movements to and from the Strait of Gibraltar. Among all the encounters I had with orcas, no doubt the one I can't easily forget occurred in June a few years ago", Alfredo told me.

"On that day, I left the harbour with two colleagues on board a small tour vessel from Ocean Vibes Algarve, a tourism and research company I started with a friend. Our goal was to collect cetacean sighting data. It was around five in the afternoon, and soon after getting into the open ocean, we noticed a pod of orcas hunting tuna very close to shore, near Isla Deserta".

"I was not so lucky this morning," I said, "having gone back and forth along the coast of Andalusia without being able to find any orca!"

"Ah, but wait till you hear my story: maybe you weren't unlucky, after all" said Alfredo. He recounted his concern once he had a good look at the orcas and realised that they were the notorious Gladises from the Iberian population, two younger individuals named Gladis Blanca and Gladis Negra. Over 6 months in the year before, these orcas had been involved in a series of rowdy interactions with various types of boats, most of them sailing vessels. These incidents had never caused a boat to sink until now but, in many cases, damaged their rudders, thereby disabling the vessels' ability to steer.

"It took us a while to confirm our suspicion that this tuna-hunting pod included these two characters. We stopped the boat dead in the water less than one mile off the inlet and remained there to watch the orcas as they engaged in hunting. We could see them chasing the tunas from Isla Deserta back to the beach of Farol, another islet, where they seemed to slow down, presumably busy finishing up a tuna they had managed to catch. We saw them slowly swimming towards the sea while passing a dead tuna back and forth between them. In one of these forays, a younger orca, possibly one of the two Gladises, swam under our boat, and we could see it quite well as it carried the remains of a tuna carcass in her mouth".

"The hunting behaviour lasted for a while, always with the same pattern: the orcas would swim in fast pursuit of the tuna westward, towards Cape Santa Maria on Deserta, and then they would rush back eastwards towards Farol, where they slowed down swimming away with a tuna carcass in their mouths".

"After about an hour of this show, we noticed that the orcas had started to swim slowly along the west side of Deserta, and we thought they were getting ready to leave the area. We fired up the engine and headed back into the inlet towards the harbour. Then we noticed two younger orcas fast approaching the stern of our vessel: one swimming belly up, the other in a normal position coming from the port side. We were very close to an inlet called Ria Formosa when the orca from the port side rammed our boat with a loud thud that reverberated along the fibreglass hull. We could feel the power of that animal! Luckily, it was touch and go: the boat was pushed slightly off her course and suffered no damage. Then the orcas dived and joined their pod as we headed back into the lagoon across the inlet".

"I had met this pod before without any incidents or interactions. But on this day, they decided to give us a touch – as they did with many other vessels before and after that episode. This sighting was remarkable because of that surprise push and because we had the chance to observe their hunting dynamics so close to the islands, marvelling at the speed they could reach during the chase. Seeing them passing under our boat carrying the dead tuna was fascinating. I must admit that I was glad that our interaction happened so close to the inlet and that they stopped right there because the feeling of an orca hitting a 600 kg boat pumped our adrenaline levels to a high in a matter of seconds".

On hearing Alfredo's amazing and somewhat disquieting story, my first reaction was to agree that indeed it had been a good thing I had not found the orcas that day, potentially exposing the wooden hull of beloved *Pontoporia* to the whim of a couple of impish orca teenagers. However, Alfredo's story did not surprise me, as I knew of several episodes of orcas making contact with vessels. First and foremost was the story of three fellow sailors and close friends—Claudio Cuoghi, Giorgio Di Mola, and the late Francesco Longanesi Cattani—who were crewing aboard the Italian sailing yacht *Guia III*, hit and sunk by an orca in the tropical west Atlantic in 1976 during the last leg of the Atlantic Triangle race, between Rio de Janeiro and Portsmouth.[6]

Half a century later, the event is understandably still vivid in the memory of one of the protagonists. After crewing on several oceanic racers in his early 20 s, Claudio spent half his life in the United States as a businessman. Now happily retired, he is my neighbour on the island of Patmos. I knew the story well because I was beginning my career as a student of cetaceans then, and the news of an orca hitting and sinking a boat was sensational. Claudio enjoyed repeating his story, and I had heard it recently while we savoured a chilled Assyrtiko wine.

"We were sailing in fair weather conditions, having left the coast of Brazil in our wake long before. The trade winds were blowing at 25 knots, and *Guia* was sailing at a brisk 7-8 knots", he told me.

"I was at the helm when the boat was hit, and *Guia* stopped dead in the water", Claudio recounted. "We immediately noticed that water had started gushing inside the boat from a large hole, about 40 by 30 centimetres near the bow, through the hull's marine plywood planking, which was 20 millimetres thick".

"I was the first to notice then that there were four to five big orcas calmly circling the boat, one of them very close to the bow, at almost touching distance", said Claudio.

"Did you actually see an orca hit the vessel?" I asked him.

"No, but what else could have happened? It was full daylight, and any debris, like a floating ship container or a tree trunk, would have been visible. But all we could see were the orcas swimming around the sinking boat".

The crew threw the life raft in the water and clambered aboard, some having jumped into the sea in the process. Meanwhile, the orcas did not visibly react to the men in the water or those aboard the life raft. *Guia* sank 15 min later, and the orcas disappeared at about the same time. The incident did not end in tragedy only because the survivors had the extraordinarily good luck of being rescued just hours later by a Greek merchant ship that was tens of miles off her course.[7]

Why would an orca go through the trouble of sinking a sailing vessel in the middle of the ocean? And what is inducing the mischievous Gladises to

---

[6] Di Mola G. 1978. SOS il Guia affonda. SM Silvio Mursia e C editore, Milano. 160 p.

[7] A Wikipedia entry exists with an extensive list of orcas' incidents with humans: https://tinyurl.com/5n8nj76h. The site lists incidents into two groups: with wild orcas and with captive orcas.

repeatedly harass hapless sailors who happened to cross their path off the coast of the Iberian Peninsula? The most commonsense thought that comes to my mind is that these boisterous young giants had simply decided to have some fun. And why not try to push their luck with some of those pesky, floating human contraptions filling their oceanic home? In recent months, the phenomenon has continued to grow. At the time of this writing, more than 500 episodes of orca-vessel interactions—including a few instances in which the boats were made to sink—have been recorded since the year 2020 along the coastal waters of the Iberian Peninsula, between the Strait of Gibraltar and Galicia. This situation has understandably created considerable excitement not only among boaters but also in the media, with the expected production of considerable sensationalist garbage. However, it has also provided an opportunity for a group of Spanish researchers—GT Atlantic Orca[8]—to set up an excellent information platform on the phenomenon. Quite reasonable voices are also being heard to describe the situation and to try to understand its causes, including an intriguing exploration of the greater context in which these intelligent animals are forced to cope with the human invasion of their world, told by author Sarat Colling.[9]

I ended the day sitting at the table of a small restaurant facing the harbour of Tarifa, in full sight of the top of *Pontoporia's* mast, slowly swaying with the ocean swell that rolled into the port, dampened by the breakwater. I took total comfort from the morning's unsuccessful orca expedition by enjoying a delicious *gazpacho cortijero*, a variety of the celebrated Andalusian cold tomato soup made even more interesting by the addition of toasted crushed almonds and peeled grapes. Nor could I abstain, with the thought of the imminent gastronomic deprivation imposed by several more days at sea, from gorging on an excessive collection of tapas such as *croquetas, pan con tomate, patatas bravas, tortillas españolas*, and *pimientos de Padrón*.

Day 41: Tarifa, Spain, to the Alborán Sea

It was now time to take to sea again for the last leg of the journey to my final destination on the Italian Riviera, crossing the waters of the Pelagos Sanctuary for Mediterranean Marine Mammals. As I cast off from the dock around midmorning after a lazy breakfast, the summer day was still hazy and damp, but the forecast was of clear weather and calm wind. As I cleared the harbour's

---

[8] GT Atlantic Orca: https://tinyurl.com/2sxvbnez.
[9] When orcas speak: listen carefully https://tinyurl.com/59nf788m.

breakwater and came into the waters of the Strait, I could finally take advantage of the favourable current entering the Mediterranean that aided the labour of *Pontoporia*'s little engine. Soon, I had the Bay of Algeciras to port. Then, around noon, the Rock of Gibraltar—famously populated by wild macaques, the only other primates with which we share the European continent—rose on the port beam whilst the coast of Morocco, to the south, was progressively fading away in the haze. I was forging ahead into the Mediterranean again, and despite the fleetingness of my oceanic foray the day before, a pleasing sense of homecoming was pervading *Pontoporia*'s one-person crew.

I proceeded eastward into the Alborán Sea again, along the "Alborán Corridor IMMA", designated to encompass the migratory paths of sperm whales and fin whales.[10] This time, I decided to maintain a route closer to the coast of Andalusia to take advantage of the favourable direction of the northern portion of the clockwise gyres.

My course initially took me farther from land, as the coast arched near Marbella and then Malaga to the north. When I had Malaga abeam to port, about 30 miles off the Spanish coast, I spotted a commotion on the calm surface of the water approximately one mile ahead, a sure sign that some large marine creature had appeared there from the depths. I had just entered the "Alborán Deep IMMA", established because of its relevance to all the species of deep-diving cetaceans found in the Mediterranean, including Risso's dolphins as well as sperm, Cuvier's and pilot whales,[11] and I was hoping that the surface disturbance might indicate the presence of one of these four species. As I neared the area, it became evident that I had run into a large group of cetaceans loafing at the surface, seemingly without anything special to do except rest and interact lazily with each other. The animals' colour was dark, almost black, and most of what I could observe before seeing them through the surface was their low, rounded, rear-projecting dorsal fins: the unmistakable mark of pilot whales (Fig. 9.2).

I had not previously encountered these intriguing members of the Mediterranean fauna for a good reason: pilot whales confine themselves to the western basin, and only they know why. Again, I am at a loss regarding the inadequacy of the English language when it comes to naming cetacean species. At a minimum, I have two good reasons for complaining about the name "pilot whale": first, because pilot whales are dolphins, not whales, and second because they are no more pilots than most of their other highly social relatives.

---

[10] Alborán Corridor IMMA: https://tinyurl.com/2dccnpja.
[11] Alborán Deep IMMA: https://tinyurl.com/2s3fjdd6.

Fig. 9.2 When looking at the head profile of this spyhopping long-finned pilot whale, one can have little doubt about the reason why the French zoologist René Lesson called the genus *"Globicephala"* ("globous head") in 1828 (Illustration by Massimo Demma)

They were given the pilot name due to the pod's purported habit of following a "pilot" in the group. Pilot whales are very social animals like all members of the dolphin family. Since they live in stable groups usually composed of tens of animals, there is nothing special about the group following the lead of the older animals. Compare this with the far more appropriate name *Globicephala* (globe head), given to the pilot whale genus by the French nineteenth-century zoologist René Lesson. The genus name is reflected almost verbatim in Italian and French. At the same time, the Spanish gave them the name *calderón* (stew pot), again referring, in a show of greater imagination, to the round shape of the head.

Pilot whale societies are matriline, like those of the orcas, and stable in time, although perhaps not as stable as orca societies. Also similar to their formidable relatives, the males mate with unrelated females during temporary aggregations of multiple groups. Two species of pilot whales are found in the world's oceans.[12] The one occurring in the Mediterranean is called "long-finned pilot whale" because it has very long, scythe-like pectoral fins. The males of the species reach almost seven metres, exceeding female length by at least 1 m. Long-finned pilot whales are found in temperate and cold waters across the North Atlantic and throughout the temperate and cold latitudes of

---

[12] Pilot whales, genus *Globicephala*: https://tinyurl.com/24hxv8a2.

the entire Southern Hemisphere. The second species, tropical in distribution, is called "short-finned pilot whale" and is a bit smaller, with shorter pectoral fins as its name implies. The ranges of the two species make it seem as though they have agreed on how to divide the oceans between themselves, except for the northern part of the North Pacific where neither species is found. Pilot whales, particularly the long-finned species, are often seen peacefully swimming along and minding their own business. However, their peaceful demeanour belies a fierce side of their character, as they have been seen harassing, for unknown reasons, the much larger sperm whales and even the famously fearsome orcas.

The western Mediterranean hosts two distinct subpopulations of long-finned pilot whales, neither of which is having a swell time. Subtle genetic differences distinguish the two subpopulations, and both differ from their North Atlantic conspecifics. The Inner Mediterranean subpopulation is the larger of the two, consisting of about 2500 individuals spread between the eastern Alborán Sea and the Ligurian Sea and assessed as Endangered on the IUCN Red List mostly due to a major drop in numbers caused by a morbillivirus infection in 2006 and 2007. High tissue levels of man-made contaminants, disturbance from maritime traffic, and damaging interactions with fisheries further compromise its survival.[13] The tiny Strait of Gibraltar subpopulation, limited to the Strait and the western Alborán Sea waters, is ten times smaller and is assessed as Critically Endangered.[14]

The pod I encountered that day was swimming at the boundary between the two subpopulations, so I had no idea which one I was seeing, although probability suggested it was the Inner Mediterranean one. At about 100 m from the group, I turned off the engine, and *Pontoporia* glided for a minute or so on her headway before stopping completely. The silence was broken only by the sound of the breathing cetaceans, at least 30 strong, as they slowly moved around, minding their own business. The animals were obviously in resting mode, likely having just surfaced after a long, deep dive in search of their squid quarry.

Eventually, the presence of *Pontoporia* seemed to elicit in the whales a mild curiosity as a few of them approached and lazily swam around the boat. Some were huge, half as long as my vessel, and certainly adult males. Many smaller animals also appeared to be females, as they were accompanied closely by young calves swimming with them in tight coordination.

---

[13] The long-finned pilot whale, *Globicephala melas*, Inner Mediterranean subpopulation, is assessed as Endangered in IUCN's Red List: https://tinyurl.com/44sbzx3j.

[14] The long-finned pilot whale, *Globicephala melas*, Strait of Gibraltar subpopulation, is assessed as Critically Endangered in IUCN's Red List: https://tinyurl.com/yb2enfyf.

At one point, a large male swam up next to *Pontoporia*'s cockpit and emerged vertically, raising his bulbous head high over the water, a manoeuvre technically known as a spyhop, a common cetacean behaviour and one often performed by pilot whales. The large head emerged slowly, less than 5 m from me, and the animal's eye made contact with mine for a few seconds. I wish I knew what he was thinking when he looked at me. Interest or aversion? Friendship or contempt? More likely, a passing mild curiosity soon to be forgotten. Whatever it was, I would never know, as hard as I tried to interpret the sparkle from his eye. The absence of facial expression caused by the hard tissues of the head did not help.

I never wish to end deliberately any close contact with cetaceans. In this case, the pilot whales seemed content to loiter around *Pontoporia* forever, but I still had a long way to go. Somewhat reluctantly, I bade goodbye to the big, black dolphins and resumed my course to the northwest as the night descended upon the Alborán Sea.

My course was now converging with the line of the coast. As I neared Andalusia, I was approaching Europe's largest concentration of greenhouses—reportedly the world's largest—all clustered together just west of Almería. Occupying an area greater than 300 km² and covered by an oceanic expanse of plastic sheets, the greenhouse agglomerate sends an eery greenish light into the night sky when seen from the sea, apparently one of the weirdest human artefacts visible from space. These greenhouses, which take advantage of the region's favourable climatic conditions combined with abundant (albeit not unlimited) water extracted from the aquifer via artesian wells, provide fruit and vegetables to the whole of Europe in defiance of both seasonality and concern for sustainability. The area accounts for 38% of Spain's horticultural production, 30 times that of the average European farm, and employs tens of thousands of workers. Customers in Madrid, Paris, and Stockholm are happy to eat the strawberries produced here in January, whilst concern for the planet's future is separated from any evidence and dissolves into venial insouciance. With some apprehension, I wonder about the fate of the 300 km² of plastic sheets once the water from the wells is depleted and the greenhouses are demolished. I hope the sheets will not be thrown into the sea or incinerated to produce airborne poison.

Soon, the lights of the monstrous greenhouse cluster faded in my wake whilst those of the city of Almería appeared abeam as I proceeded eastward into the night.

Day 42: Alborán Sea

It was mid-morning when I rounded the brown cliffs of Cabo de Gata, thereby exiting the Alborán Sea and entering the vast Gulf of Vera, where the Andalusian coast yields to that of the Murcia region. It was time to alter my course a few degrees to the north towards the Balearic Islands.

The cape I had just passed is the centrepiece of an important Spanish protected area, the Cabo de Gata-Níjar Natural Park, Andalusia's largest coastal park, endowed with a beautifully wild landscape juxtaposed in stark contrast to the heavily urbanised and cultivated area it borders to the west. I remembered visiting the park by land a few years before and being impressed not only by the pristine beauty of the place but also by the highly professional and effective management of the protected area. Spain should be credited with her resolve to do things right as far as her parks on land and at sea are concerned, or at least better than most other nations, particularly compared to the dismal general condition of MPA management in the Mediterranean.

MPAs are one of the most important tools—perhaps the most important one—in the toolbox of the marine conservationist. As I noted when sailing around the Egadi Islands a few days ago, MPAs are also an admission of the human inability to treat the sea with respect, thereby requiring us to carve out tiny pieces of the ocean that we subject to special regulations in an often-clumsy attempt to ease our conscience. Establishing a protected area, particularly a protected area covering a marine surface, can be a laborious and costly process, requiring, in most cases, excruciating negotiations with the local stakeholders. And this already means starting with the left foot, because, in a normal world, it should be the stakeholders themselves the ones requesting that a protected area be established to preserve their space.

Once protection for an area is considered, careful scientific assessments are performed to justify the MPA designation, culminating in often uncertain passages through the relevant national parliament before a law is passed, and the MPA is sanctioned—on paper. The whole process can take many years. Then management must kick in because until actions are implemented to protect an area concretely and effectively, the park remains a so-called paper park.[15] Sometimes, however, conservation objectives at sea can be reached without having to go through the rigmarole of formally establishing a protected area. When protection is sought from a single, specific threat, and that threat can be addressed with a specific legal tool, the goal can be attained much faster and less painfully. Here, in the waters off Cabo de Gata, was the clearest example of the latter alternative, harkening back to 2006.

The Mediterranean Sea is shaped such that the main shipping lane connecting the Strait of Gibraltar to the opening of the Suez Canal in Port Said, lies a few miles offshore of Cabo de Gata. Ship traffic impacts marine life through pollution, disturbance, and noise. Ships can also hit and kill larger animals such as cetaceans, adding a significant cause of mortality to the plight of these already beleaguered mammals, as I had noted when sailing across the Hellenic Trench weeks before. In 2005, the potentially damaging impact of shipping on bottlenose dolphins and loggerhead turtles in the coastal waters off Cabo de Gata was brought to the attention of the International Maritime Organisation by the Spanish government, with a request to move the shipping lane seaward by 20 miles, thereby diverting further offshore the tens of thousands of vessels that pass the cape every year on their way to or from Gibraltar. This change came into force in 2006 and has minimally increased the distance travelled by transiting ships. However, this measure has effectively protected marine life in the coastal waters off Cabo de Gata from the pressure of shipping without the need for a convoluted legal process to establish blanket marine protection there.

Pontoporia was now proceeding briskly under sail, taking advantage of the pleasant sea breeze that had started to blow towards the coast around midday. The Spanish mainland was getting farther and farther away with the passing hours as I sailed to the northeast, intending to hug the southern coast of the Balearic Islands. The day passed, followed by night, in a progressively less trafficked sea as I left the main trans-Mediterranean maritime shipping lane astern.

Days 43 and 44: Western Mediterranean Sea

Hours passed into days without events worth reporting here as I approached the end of my journey. Ibiza and Formentera, the smallest of the Balearic Islands, had appeared above the horizon to port around noon on day 43, less

---

[15] Relano V., Pauly D. 2023. The 'Paper Park Index': Evaluating Marine Protected Area effectiveness through a global study of stakeholder perceptions. Marine Policy. https://doi.org/10.1016/j.marpol.2023.105571.

than 20 miles to the north, and soon disappeared. I was sailing in rather deep waters, just off the islands' continental shelf. On my approach to Mallorca, the waters became progressively shallower until I entered the "Balearic Islands Shelf and Slope IMMA",[16] where I expected to run into bottlenose dolphins closer to the coast and sperm whales farther offshore, in the deeper waters.

Later that afternoon, I passed a few miles south of the island of Cabrera, the seat of the Cabrera Archipelago Maritime-Terrestrial National Park, another gem of marine protection in the Mediterranean, very well-managed by the authorities in charge. The park hosts important breeding colonies of marine birds such as storm petrels and shearwaters, including the critically endangered Balearic shearwater.[17] Moreover, in addition to preserving a representative ensemble of coastal and marine species of the western Mediterranean, its waters are visited by open seas summer visitors such as fin whales and spinetail devil rays. That evening, the coast of Mallorca passed nearby to my port side, and then, as the sun rose on the following day, Menorca, the easternmost of the Balearic Islands, came into full sight. By the afternoon, I rounded the low cape of Punta Prima, about 6 miles from Menorca's coast, sailing at the edge of the continental shelf where the sea bottom to starboard dropped off steeply, descending within a handful of miles to depths greater than 2500 m.

As the coast of Menorca was streaming by to the port side, not far from the entrance to the island's main town of Port Mahon, the vast expanse of the Sardinian-Balearic abyssal plain came into view, bordered on the north by the Gulf of Lion and on the east by the coasts of Sardinia and Corsica: a highly dynamic oceanic ecosystem full of life, and a very important one for many of the region's largest marine inhabitants. A light thermal breeze was blowing towards land, causing the sea surface to be covered with tiny ripples, when to starboard, no more than one mile offshore, my eye caught the glimpse of a big tail coming out of the water, and my heart skipped a beat. I had run into sperm whales again!

The glimpse of the whale had finally pumped some adrenaline into my bloodstream after a few days of uneventful navigation. I knew that the distant whales were sperm whales by virtue of the tell-tale sight of the tail raised high above the water at the beginning of a dive. And nobody on Earth could prevent me from approaching these whales and finding out something more about them: how many of them were there? Was it just a single large male or a social unit made of females with their young? Or both, like the group I had

---

[16] Balearic Islands Shelf and Slope IMMA: https://tinyurl.com/45e6tfp7.
[17] The Balearic shearwater, *Puffinus mauretanicus*, is assessed in IUCN's Red List as Critically Endangered: https://tinyurl.com/yc7rwysj.

met south of Crete? What is he, or what are they, doing? These are all the usual questions that come to mind when sighting whales.

I detoured towards the spot where I had seen the tail disappearing in the water, and minutes later, *Pontoporia* was floating motionless in the location of the dive: engine off, perfect silence, a soft tinnitus slowly dissipating in my ears after so many hours of engine noise. The excitement of spotting the whales and waiting for them to appear from the deep was fresh and spellbinding as if it were for the first time in my life. For almost a half-hour, nothing happened, and I began to wonder whether that whale tail flicking up had been just a figment of my imagination. To dispel that disturbing thought, I turned on the hydrophone, and sure enough, the signature hammer-like sounds made by diving sperm whales when hunting were coming out from the speakers loud and clear. However, the sperm whale I was listening to could have been miles away, in which case I would have had no chance of seeing it when it surfaced. Sperm whales engaged in hunting deep-water squids can remain submerged for over 1 h, although dives last, on average, about half that. I was keen on having a close encounter with these whales, but I was also mindful of the 290 nautical miles yet to cover before completing my journey.

As I was dealing with the internal struggle of whether to stay or go, I heard the loud noise of the whales drawing their first powerful breaths as they emerged not far from *Pontoporia*. Waiting there had been the winning choice, and in minutes, I was surrounded by several surfacing sperm whales. As the pod rested after the long dive, I could count seven of them, all medium-sized. By comparing their size with the length of *Pontoporia,* I guessed their lengths as being between 10 and 12 m. The absence of young in the group indicated that these whales were most likely a bunch of male bachelors, possibly aged between 10 and 15 years. I was not surprised to find sperm whales in this place because the deep waters to the south of the Baleares are among the most favoured areas in the Mediterranean Sea by the species: a prime squid hunting ground, and less trafficked than most places, certainly less than the main shipping lane further to the south.

It was now late afternoon, and the whales around *Pontoporia* had been quietly resting for a while. And then, as if on cue, one by one, they made a strong blow, arched their backs high over the waterline, lifted their large tails to push themselves vertically towards the depths in a flourish of power and intention, and within the span of a few seconds all had disappeared from sight, likely in search of the second course of their dinner. More modestly, I unwrapped and ate the sad remains of a tortilla I had bought in Tarifa, after which I lifted the sails to take advantage of the gentle breeze blowing towards Menorca. And off I headed towards the Ligurian Sea as the night descended and an extra-large full moon rose above the horizon.

Day 45: Western Mediterranean Sea

The breeze had died some distance from land during the night, so I lowered the sails and proceeded under the engine across the calm waters that the moon had transformed into liquid silver. At dawn, I was in the open sea about midway between Sardinia and the coast of Catalunya. I had just entered the "North West Mediterranean Sea, Slope, and Canyon System IMMA",[18] established there in view of the important habitat it contains for fin and sperm whales, Risso's dolphins and Cuvier's beaked whales, and more generally for all the cetacean species regularly occurring in the Mediterranean. This IMMA is one of the most important areas for cetaceans that I have crossed since departing from Venice, one with world-class relevance for marine mammals. Given the area's importance for large cetaceans such as fin and sperm whales, the International Maritime Organisation has recently established here a Particularly Sensitive Sea Area (PSSA) to stimulate the adoption by the national authorities of rules to mitigate the impact of maritime traffic on the whales.

But this area is also one of the stormiest in the Mediterranean, as the notorious Gulf of Lion is a portion of the Mediterranean on a par with the Aegean Sea with regard to the frequency and intensity of its winds. Thankfully, I must have been in the good graces of the god of winds, as the sea surface before me was flat as a mirror. I was relieved to have been spared a crossing under the whip of the Mistral, the fierce northwesterly wind that gathers speed along the valley of the Rhone River before irrupting with its might into the Gulf of Lion, able to wreak havoc on the unlucky mariners venturing in the wrong time across this body of water.

Relief about the weather had left room for much more exciting thoughts. I was full of anticipation for the chance to finally meet the flagship of all Mediterranean cetaceans: the mighty and gentle fin whale.

How could I ever forget my first encounter with a whale? It was indeed a fin whale, sighted not too far from where I was that day. I was in my late teens, crewing on a sailboat racing in the Giraglia, a well-known Mediterranean offshore regatta so-called because it involves reaching and rounding a small rock with that name off the northern tip of Corsica before returning to the continent. On that day, the effect of an early summer anticyclone had caused the wind to be absent except for capricious afternoon breezes limited to the coastal zone. As a result, our boat was stuck in a dead calm in front of the Giraglia for hours. Turning on the engine was out of the question given that we were racing, or at least we were supposed to be.

---

[18] North West Mediterranean Sea, Slope and Canyon System IMMA: https://tinyurl.com/m6xsrcc9.

The conditions for a session of sea watching were perfect, something that seemed much more fun than struggling to get the boat to move when moving was no more than wishful thinking. And there was plenty to watch. The sea surface was strewn with a multitude of blue button jellies, tiny round critters known to science as *Porpita porpita*, distantly related to jellyfish, that form into elegant bluish mini-colonies of floating hydroids. Attracted by the bonanza, scores of mid-sized ocean sunfishes[19] were bobbing at the surface as far as the eye could see, clumsily feasting on the jellies, and the sight abundantly compensated me for the anticlimactic racing conditions. And then, literally out of the blue, a fin whale erupted in front of our boat with a startlingly explosive blow. Its appearance was so sudden and unexpected, and its size so unbelievably immense, that the image was etched in my neurons to remain there until the end of my days, accompanied by a sense of awe and affection for such a huge and yet so unobtrusive a giant.

Back to the present, aboard *Pontoporia*, even before seeing them I knew that fin whales were around. It was not just that I was in the right place, at the right time of year, with sighting conditions that could not have been more favourable on that calm day. It was a certainty from deep inside, from some primaeval hunter layer embedded in my cerebral cortex. My eyes were scanning the horizon, patrolling right and left along that most distant narrow line just below the separation between sky and sea, encompassing the broadest possible sea surface. I knew from experience that it was there that whales would first appear if any of them were around. It was a matter of habit, an automatic behaviour no longer requiring the command of intention: the eyes go back and forth and as soon as an anomaly in the pattern is detected—the minimal deviation from linearity or immobility—only then is conscious attention evoked. So it is that an innate tension binds predator with prey and allows the survival of either depending on circumstances, requiring the life-or-death imperative of fine-tuned attention to the surroundings. The situation in my case was different because it involved only me in the role of the hunter. The hunted whales could not care less whether I sighted them or not because being predated in the Mediterranean is the least of these giants' worries. Anyway, my predatorial intent was limited to locating and approaching my quarry at a safe distance to enjoy the sight without causing disturbance.

And, indeed, on that day, I did not have to wait long. There it was, that tiny dark speck moving swiftly just below the horizon line, right ahead of *Pontoporia*, a mile or perhaps two ahead, slightly to starboard. It is counterintuitive that finding fin whales in a wide marine expanse requires knowledge

---

[19] Ocean sunfish, *Mola mola*: https://tinyurl.com/2kvzzc2r.

and determination. They are the second largest animal on Earth. Yet, unless you are intently looking for them, you know what to look for, and you wish them to appear with all the strength of your most intense resolve, you will miss them and sail across their habitat without the faintest notion that you had sailed amongst whales. I am convinced that the difficulty of seeing whales is one of the main reasons why so few people today are aware of their common presence in the Mediterranean.

Confirmation that that minuscule irregularity below the horizon, barely stimulating a few cones in my retina, was indeed a fin whale came from its behaviour. The tiny shape eventually slowed, becoming thicker before disappearing as the whale arched its back high above the waterline and dove. In doing so, unlike sperm whales, it dove without showing the tail, thus revealing its fin whale identity. As the first whale disappeared underwater, similar silhouettes were busily moving just under the horizon right and left. There were perhaps half a dozen fin whales out in the distance. Each was doing its own thing, most likely searching for the best spot to dive and gorge on krill, the tiny planktonic shrimp that live in swarms hundreds of metres below the sea surface, and that draw the whales to this part of the Mediterranean at this time of year (Fig. 9.3).

I admit that, amongst all cetaceans, I have a soft spot for fin whales. They are immense. With a maximum length greater than 20 m, they are the largest animal ever, second only to their close relatives, the blue whales. Significantly bigger than sperm whales, unlike sperm whales they are beautifully streamlined, as if designed by a naval architect with a special talent for aesthetics. When a fin whale materialises in front of you out of the deep, you cannot but think that something unearthly has appeared, if only because of its sheer size, as if your senses rebelled to admit that anything that large can be alive. And yet, their immensity is not monstrous, partly because they are so beautiful and partly because of their discreet, unobtrusive behaviour, so wary and careful to avoid being near humans. They swim by, giving, at best, a sense of being indifferent to your presence, with such indifference sometimes bordering on that sort of aloofness that befits royalty. Fin whales rarely display to humans the curiosity, or even the confidence, that other members of the cetacean tribe so generously dispense, like dolphins for example.

Aside from their demeanour around humans, their migratory habits make fin whales a particularly intriguing cetacean. They certainly warrant inclusion among the great migratory whales, such as the grey, humpback, right and blue whales; however, with respect to migration, fin whales are different. The general pattern of whale migration across the world's oceans involves a poleward movement during summer to take advantage of the warm season's long

Fig. 9.3 Millions of years spent travelling and chasing krill and small fishes have allowed fin whales to evolve extreme body streamlining, quite well adapted to the migratory habits of this ocean voyager (Illustration by Massimo Demma)

daylight hours to gorge on zooplankton, and an opposite movement towards the tropics in winter to mate and give birth in an environment that is milder and more favourable to newborns. The migrations of most baleen whale species follow this rigidly seasonal pattern, which includes, not by chance, a gestation period of 12 months.

Fin whales do not, however, seem to follow such a predictable routine. Studies of fin whales tagged with electronic transmitters have allowed scientists to track them across vast oceanic expanses via satellite signals and demonstrated fin whales' propensity for doing their own special, often individualist thing. Defining the migratory pattern of fin whales at a global scale remains a challenge; instead, their movements could best be described as a mix between the classic polar-to-tropics-to-polar model, applicable to large whales in general, and major departures from that model, depending on a variety of circumstances. Fin whales thus appear to be a combination of typical great migrators and oceanic nomads, capable of taking advantage of small-scale, short-lived productivity phenomena occurring in their watery medium to find their food, which they seem to locate with fine ecological clairvoyance.[20]

---

[20] Geijer C.K.A., Notarbartolo di Sciara G., Panigada S. 2016. Mysticete migration revisited: are Mediterranean fin whales an anomaly? Mammal Review 46(4):284–296. https://doi.org/10.1111/mam.12069.

Fin whales are at home in all the world's oceans, ranging from the poles to the tropics, and they are also represented by separate populations in many smaller seas, including the Mediterranean—where there are about 3000 of them.[21] This is quite a good-sized population of fin whales for such a small sea, and even scientists were quite astonished when that order of magnitude related to the Mediterranean fin whale population first appeared based on earlier surveys in the late 1990s. Readers might wonder how one gets to estimate the population numbers of whales scattered over areas hundreds of thousands of square kilometres wide, and this is, indeed, a legitimate question. The method that has gained most traction today, to the point of being universally used, is that of the "line-transect" survey.[22] Researchers first design a grid over a map of the area to be surveyed and follow the tracks so defined, either by boat or—more commonly—by plane, sampling sightings along those tracks. Sightings are then analysed based on many variables, including the number of animals seen per sighting and the distance of the sighted animals from the track. These data lend themselves to a calculation of the mean density of whales within the narrow strip of surface visible on both sides of the track. The total whale population is then derived by extrapolating the density value in the strip to the overall area. All these calculations today are made by dedicated software that also provides the uncertainty margins. The latest estimate of the Mediterranean fin whale population mentioned above was made in 2018 by an initiative of the "Agreement on the Conservation of Cetaceans in the Black Sea, Mediterranean Sea and Contiguous Atlantic Area" (ACCOBAMS).[23] I expect things to change significantly in the future, with sightings eventually collected automatically by video from drones, or even better, through satellite imagery. At this moment in time, however, sending real people in the air day after day to cover the surveyed area properly is still cheaper than acquiring satellite images with the proper resolution.

But are fin whales really a distinct population in the Mediterranean? Again, the picture still defies clear interpretation, as even in this small portion of the world's oceans it is a challenge to figure out what fin whales really do. I still remember the day, back in 1994, when we first managed to collect tiny skin biopsies from free-ranging fin whales in what was later to become the Pelagos

[21] The fin whale, *Balaenoptera physalus*, Mediterranean subpopulation, is assessed as Endangered: https://tinyurl.com/bdyxj737.

[22] Hammond P.S. 1986. Line transect sampling of dolphin populations. Pp. 251–280 in M.M. Bryden and R. Harrison (eds.), Research on dolphins. Clarendon Press, Oxford. 478 pp.

[23] Panigada S., Pierantonio N., Araujo H., David L., Di-Meglio N., Doremus G., Gonzalvo J., Holcer D., Laran S., Lauriano G., Paiu R.M., Perri M., Popov D., Ridoux V., Vazquez J.A., Cañadas A. 2024. The ACCOBAMS survey initiative: the first synoptic assessment of cetacean abundance in the Mediterranean Sea through aerial surveys. Frontiers in Marine Science https://doi.org/10.3389/fmars.2023.1270513

Sanctuary for Mediterranean Marine Mammals. I was keen to provide our colleague Martine Bérubé from the University of Copenhagen the samples needed for genetic analyses to determine whether "our" whales were part of an Atlantic population seasonally migrating in and out of the Mediterranean or instead a separate tribe residing permanently in the Mare Nostrum. The second possibility would obviously have had significant conservation implications.

At a marine mammal conference the year before in Galveston, colleagues C. Scott Baker and Mason Weinrich gave me the prototype of a small steel tip to collect whale skin biopsies. The tip was made to be fitted onto a crossbow bolt, which would be projected into the back of a fin whale as it emerged from the water to collect a small piece of skin and blubber for scientific analyses. Back home, I had the tip replicated multiple times by a local blacksmith and then fit the tips onto modified crossbow bolts equipped with polystyrene foam for flotation. The whole contraption had a horrendous makeshift appearance and added significant anxiety to the undertaking, especially given that I had to battle my instinctive aversion to shooting a bolt into the flank of a revered creature.

But shoot the bolt I did, and it all went incredibly well: the bolt bounced from the back of the whale (who did not even flinch) and floated calmly in the whale's wake for us to collect it. And—marvel!—the tip contained a wonderful little plug of skin and blubber, emanating a faint fishy smell. That skin biopsy became the first of many, particularly thanks to Simone Panigada and Margherita Zanardelli, two former students of mine, now married, who continued to study fin whales in the Pelagos Sanctuary long after I had to interrupt my fieldwork to enjoy the delights of Capitoline bureaucracy. Simone, today the head of the Tethys Research Institute, has become a well-known sharpshooter, enlisted to collaborate in a dizzying variety of research projects from Mexico to Antarctica to biopsy and implant satellite tags on whales. Margherita is the diligent keeper of the Mediterranean fin whale catalogue, which contains by now the portraits of more than 500 individual animals.[24]

We used the biopsy technique sparingly to minimise bothering the whales. Still, we have collected enough samples to provide sufficient material for our geneticist colleagues to dispel ambiguities about the population separation between the Mediterranean and Atlantic whales.[25] The added benefits of the

[24] Zanardelli M., Airoldi S., Bérubé M., Borsani J.F., Di-Meglio N., Gannier A., Hammond P.S., Jahoda M., Lauriano G., Notarbartolo di Sciara G., Panigada S. 2011. Long-term photo-identification study of fin whales in the Pelagos Sanctuary (NW Mediterranean) as a baseline for targeted conservation and mitigation measures. Aquatic Conservation: Marine and Freshwater Ecosystems. https://doi.org/10.1002/aqc.3865.
[25] Bérubé M., Aguilar A., Dendanto D., Larsen F., Notarbartolo di Sciara G., Sears R., Sigurjonsson J., Urban J., Palsbøll P.J. 1998. Population genetic structure of North Atlantic, Mediterranean Sea and Sea

technique, which we have significantly improved over time, include the ability to genetically determine the sex of the biopsied whale and to analyse the small piece of blubber chemically in search of contaminants. Based on our biopsies, our colleagues Cristina Fossi and Letizia Marsili from the University of Siena have derived a fundamental understanding of the whales' contaminant load and their potential susceptibility to consequential intoxication.[26]

So, fin whales in the Mediterranean are genetically distinct from Atlantic fin whales, indicating that the two populations are reproductively separate. But are they really separate? No, they are not because, as we later discovered, Mediterranean fin whales, or at least some of them, swim from the Mediterranean into the Atlantic, where they have the opportunity, at least in theory, to mingle and even interbreed with their oceanic relatives. Complicating matters further is the recent evidence that a different population of North Atlantic fin whales seasonally encroaches into the westernmost part of the Mediterranean. Determining that these Atlantic invaders are different from the Mediterranean whales was achieved not on genetic grounds but on the basis of their different dialects, appearing from analyses of their low-frequency vocalisations.[27] Challenged by all these conflicting bits of information, one thing we know for sure: the simple narrative of a Mediterranean population of fin whales living in isolation from a contiguous population in the Northeast Atlantic Ocean no longer holds true. Essentially, there is nothing straightforward as far as fin whale ecology is concerned.

The summertime aggregation of whales in the northwest portion of the Mediterranean basin, i.e. right where I was sailing that day, is one of the most prominent features of the species' movements in this part of the world. Fin whales patrol their environment in a predictable seasonal pattern to sustain their gigantic bodies and, thanks to the area's peculiar oceanographic conditions, the northwest part of the basin is one of the few places where krill is available during summer in relevant amounts. I noted earlier how the productivity of Mediterranean waters is generally poor compared with the world average, presenting a problem for huge fin whales that need a large prey biomass to survive.

---

of Cortez fin whales, *Balaenoptera physalus* (Linnaeus 1758): analysis of mitochondrial and nuclear loci. Molecular Ecology 7(5):585–599. https://doi.org/10.1046/j.1365-294x.1998.00359.x.

[26] Marsili L., Fossi M.C., Notarbartolo di Sciara G., Zanardelli M., Nani B., Panigada S., Focardi S. 1998. Relationship between organochlorine contaminants and mixed function oxidase activity in skin biopsy specimens of Mediterranean fin whales (*Balaenoptera physalus*). Chemosphere 37(8):1501–1510.

[27] Notarbartolo di Sciara G., Castellote M., Druon J.N., Panigada S. 2016. Fin whales, *Balaenoptera physalus*: at home in a changing Mediterranean Sea? In: G. Notarbartolo di Sciara, M. Podestà, B.E. Curry (Editors), Mediterranean marine mammal ecology and conservation. Advances in Marine Biology 75:75–101. https://doi.org/10.1016/bs.amb.2016.08.002.

In winter, the colder air lowers the temperature of the surface water, which becomes heavier and sinks, aided by the frequent winter winds that further enhance vertical mixing. This is conducive to the onset of algal blooms and to establishing the conditions to support a decent trophic web. However, during summer, when the warm atmosphere heats up the surface layer, this process stops, and the opposite occurs: the warmer temperatures cause the surface water to be lighter than the layers below, which prevents mixing and the exposure of nutrient-rich deeper water to sunlight; a lower frequency of strong winds in the warmer season further compounds this condition. Thus, the nutrients remain deep, away from sunlight, photosynthesis is curtailed, and no trophic web is triggered. This situation is problematic for Mediterranean fin whales.

Fortunately, however, there are exceptions, as there always are in nature. Pockets of higher productivity can be found in the Mediterranean, where deep, nutrient-rich waters are made to rise to the open sea surface. Such small areas of productivity appear and disappear here and there throughout the basin in a mosaic of often ephemeral conditions depending on the movements of water masses, the morphology of the sea bottom, and localised meteorological phenomena. Fin whales, in their remarkable ecological wisdom, know how to exploit such short-lived idiosyncrasies with uncanny geographical memory, and they move about to intercept their food pockets in the right place and at the right time.[28]

The largest and most conspicuous exception to the rule which determines the summertime impoverishment of Mediterranean open sea surface waters, includes the western Ligurian Sea, the Corsican Sea, and the Provencal Basin off southern France. The area's importance for fin whales, and indeed for many other large marine predators, has justified the identification here of the "North West Mediterranean Sea, Slope and Canyon System IMMA" mentioned earlier. Upwelling caused by water circulation, invigorated by winds such as the Mistral, creates the optimal conditions for phytoplankton productivity in the open sea lasting during the warm months, in stark contrast to the surrounding areas where the water stratification transforms the summer sea into a liquid desert. And here is where fin whales flock, likely from all around the Mediterranean, to feast on the massive concentration of zooplanktonic

---

[28] Druon J.N., Panigada S., David L., Gannier A., Mayol P., Arcangeli A., Cañadas A., Laran S., Di Méglio N., Gauffier P. 2012. Potential feeding habitat of fin whales in the western Mediterranean Sea: an environmental niche model. Marine Ecology Progress Series 464:289–306. https://doi.org/10.3354/meps09810.

biomass. To use a metaphor, we could say that the whales converge here because this is the only whale restaurant open at this time of year.

Come autumn, the combined effects of surface cooling and high-energy weather systems becoming more frequent stir up the stratified surface layers throughout the Mediterranean. The consequent mixing of the layers causes conditions for increased primary productivity to expand, and ultimately, more whale restaurants open everywhere, relieving the whales from being confined to the northwest Mediterranean. This does not mean that all the whales necessarily leave the area, and a few remain there throughout the year.

An understanding of how fin whales manage their existence in a food-poor marine environment—and of how food availability is a limiting factor for the whales, governing their movements in such precise and predictable ways—leads to an obvious question: why do these whales remain confined to the Mediterranean if feeding conditions here are so ungenerous? What keeps these hungry giants within the Great Sea's narrow boundaries? Why do not they move into the Atlantic, where food is abundant, and stay there instead of returning through the Strait of Gibraltar? There must be something good inside the Mediterranean, some benefit compensating these 3000 whales for the cost of constantly dealing with food limitations. Nobody has come up yet with a clear answer to this question, but there are some ideas. For example, the notion that the Mediterranean might perhaps provide a relatively more hospitable physical habitat than the North Atlantic, with milder weather and climate and more favourable temperature ranges, combined with zero predation pressure from the orcas and, yes, from humans as well.

Indeed, one of the peculiarities of the Mediterranean Sea of special relevance to the whales is that they have never been systematically hunted here. Although the peoples of the region do not stand out as particularly caring stewards of their marine environment, we owe them at least the appreciation that they never succumbed to the greedy nonsense of embarking onto the wanton and cruel practice of whaling on an industrial scale. There was one short-lived exception at the beginning of the twentieth century when a whaling industry operated around the Strait of Gibraltar—albeit mostly on the Atlantic side—under Norwegian instigation. The hunt quickly and predictably depleted the area's whales to the point of becoming soon economically unproductive.[29] I suspect that the confusing ecological picture of fin whale

---

[29] Aguilar A., Borrell A. 2007. Open-boat whaling on the straits of Gibraltar ground and adjacent waters. Marine Mammal Science 23(2):322–342. https://doi.org/10.1111/j.1748-7692.2007.00111.x.

movement around Gibraltar, briefly described above, might be somewhat connected to the disarray caused to the populations by the whaling activities that occurred there one century ago.

On a global scale, twentieth-century industrial whaling managed to kill almost three million whales, bringing many populations and even species to near collapse, including blue, fin, and sei whales. Please pause for a second to consider the enormity of this fact: three million whales were spirited out of the ocean in less than a century to be commodified into oil and meat. It is only now that we are realising, besides the obvious welfare and ethical considerations, how such massive subtraction of the whale element from oceanic ecosystems, which continues to this day in Norway, Japan, and Iceland, impacted not only the survival of the hunted species, and not only the ecological balance of the involved ecosystems, but also the ability of the oceans to slow climate change, given the recently discovered ability of whales to facilitate scrubbing of carbon dioxide from the oceanic ecosystems.[30]

Another potential element making the Mediterranean a desirable home for fin whales might be the sea's small size, allowing the whole population to remain connected acoustically through the long-range vocalisations that these whales are known to use to communicate with each other. In the end, it is not implausible that a complex of conditions conjured to create a favourable environment for fin whales in the Mediterranean Sea thanks to an extended, overlapping calendar of feeding and breeding opportunities, whereby these two essential needs do not have to be separated in space and time as is the case for the oceanic populations.[31]

All the above ideas are interesting and perhaps even make sense, yet another question arises. Are the conditions currently deemed favourable to fin whale survival in the Mediterranean compatible with the sweeping environmental changes caused here in recent decades by human actions? Maritime traffic intensity keeps growing, and vessels travel faster and faster, increasing the risk of deadly collisions with the whales. Furthermore, thousands of propellers of all sizes and types spinning simultaneously at any time in the small basin create a low-frequency hum filling the confined underwater Mediterranean soundscape. This makes it more difficult for the whales to communicate over

---

[30] Pearson H.C., Savoca M.S., Costa D.P., Lomas M.W., Molina R., Pershing A.J., Smith C.R., Villaseñor-Derbez J.C., Wing S.R., Roman J. 2023. Whales in the carbon cycle: can recovery remove carbon dioxide? Trends in Ecology and Evolution. https://doi.org/10.1016/j.tree.2022.10.012.

[31] Notarbartolo di Sciara G., Castellote M., Druon J.N., Panigada S. 2016. Fin whales, *Balaenoptera physalus*: at home in a changing Mediterranean Sea? In: G. Notarbartolo di Sciara, M. Podestà, B.E. Curry (Editors), Mediterranean marine mammal ecology and conservation. Advances in Marine Biology 75:75–101. https://doi.org/10.1016/bs.amb.2016.08.002.

large distances, decreasing the potential for sound-mediated socialisation. And what about the danger of getting trapped in a net? Or snarled in one of those longlines, tens of miles long and armed with hooks used to catch swordfish and tuna? Or getting plastic materials down the throat, either in the form of large-sized garbage that can obstruct the oesophagus and prevent feeding, or as micro- and even nano-plastics capable of sneakily insinuating themselves into the bloodstream where they are likely to cause damage?

Even the risks caused by ship traffic, fishing, and plastic pollution pale in the face of the expected effects of climate disruption. With the mass circulation of water across the world's oceans set into motion by the seasonal balances of heat and salinity gradients, the observed warming of the Mediterranean is bound to affect water movement and, very likely, its productivity and the availability of food to the whales. Will these animals learn to adapt to the new conditions, keeping up with the rapidity of the changes in their remarkable ecological savviness, or will they, at some point, head for Gibraltar and bid us farewell for good?

Pontoporia was proceeding towards where I had first spotted the whales, and minutes later, having reached the location where I had last seen them diving, I turned the engine off and waited as I normally do in these circumstances. Fin whales are not diving champions like sperm and Cuvier's beaked whales, and their dives normally last just a handful of minutes. However, the situation in the Mediterranean is somewhat unique because here krill remains at considerable depths during the daytime, swimming towards the surface only at night. Therefore, whales diving for a snack during the day have been observed at depths of almost 500 m, which is unusual. These circumstances mean that the dive time of fin whales in the Mediterranean is longer than the species' average.

As I expected, four whales emerged, one after the other, around *Pontoporia* in the calm of high noon, their surfacing marked by the booming sound of their first blows after the dive. After that overture, the silence was so absolute that I could hear the water rippling from the whales' backs as they proceeded across the surface, seemingly oblivious to my presence. All the animals were likely adult and a good deal longer than *Pontoporia*, milling around in their seemingly effortless motion, their torpedo-shaped bodies smoothly propelled forward by imperceptible movements of their massive tails. One of them, in particular, came closer, on a parallel course to my left, showing its bright white right jaw as it cruised a few metres underwater and came in slow motion up for air until the large back broke the surface. And then came the powerful, vertical blow throwing a 5-metre-tall column of droplets up in the air, followed by quick air intake with the typical sucking noise before the head dipped under again.

There is a spellbinding beauty in the surfacing sequence of the fin whale, and I was silently imploring it, "Please, please don't dive yet, I want to watch your breathing sequence for hours on end ...". But the sequence always consists of a half-dozen surfacings at most. Minutes later, the whale arched its back high above the water to dive, and the spell was gone.

All four whales eventually ended their surface time, and again disappeared into the depths, likely to continue the lunch they had interrupted to come up for air. I still had some ground to cover before reaching port and wished to resume my course despite my reluctance to part with such a regal company. I accompanied the whales with my imagination along their underwater forays in the darkness of the deep sea for some time but soon had to turn my mind to what was coming next in my continuing journey.

Around nightfall, at about 90 miles from my destination, with quite a thrill, I crossed an imaginary line connecting mainland France just east of Toulon with the northwest tip of Sardinia, thus entering the Pelagos Sanctuary for Mediterranean Marine Mammals. Eighty-three thousand square kilometres wide, the iconic Pelagos Sanctuary is the largest MPA in the Mediterranean. I thought it was the most appropriate location for concluding my journey aboard *Pontoporia*.

The Pelagos Sanctuary was established in 1999 by a treaty among France, Italy, and Monaco to protect one of the most important areas for Mediterranean marine mammals. The treaty's signing concluded a long struggle between the imperative to conserve the marine environment and safeguarding economic interests. The story of how the Pelagos Sanctuary came into existence is representative of the omnipresent tension between environmental and socio-economic drivers that defines our times, and is thus a tale worth telling.

It all started in the early 1980s when a terribly destructive way of fishing became fashionable in the Mediterranean. The nets used in this fishery are called large-scale pelagic drifting gillnets—driftnets in short—and had spread among Mediterranean fishers to capture two of the most valuable fish species on the market: swordfish and bluefin tuna. A gillnet is called a driftnet when it is floating and not attached to the bottom like most gillnets are; it is attached only from one end to the fishing boat from where it was deployed. The net hangs vertically in the water, kept in position by a line of floats at the surface and a line of leads along its bottom, 30 m deep or even deeper: a "wall of death", as it has been aptly named. Not anchored to the sea bottom, a driftnet can be deployed anywhere in the sea regardless of water depth and is particularly suitable to catch pelagic fish in the open sea, far from the coast. Made of extremely strong synthetic polyfilament fibres, driftnets deployed in the Mediterranean were typically about 10 km long, and often much longer.

In the decades of unchecked extravagance between the 1970s and the 1990s, with the blessing of the authorities in charge, the driftnet fleet had reached the monstrous size of some 600 vessels in Italy alone. There was easy money to make in that venture, as driftnets could be set without much traditional expertise and were partly subsidised with public money. The resulting length of all the nets deployed in Italy, if attached together end-to-end, could have caught the totality of the Italian peninsula in their deadly mesh. Attracted by the abundance of swordfish and bluefin tuna in the productive waters of the Ligurian Sea, bands of driftnetters converged there from southern Italy in summer, deploying thousands of kilometres of nets.

Driftnets catch everything in their paths along with the targeted large pelagic fish, including all sorts of marine creatures and even smaller boats. Unwanted victims predictably include multitudes of cetaceans, from the larger sperm whales to the smaller striped dolphins. When an air-breathing mammal such as a whale or a dolphin gets entangled in a driftnet, its fate is sealed: prevented from reaching the surface, it eventually drowns. We know of about 230 sperm whales that were found dead during the last two decades of the twentieth century, their skins covered by the telltale signs of net entanglement. Numerous carcasses of striped dolphins, pilot whales, and even Cuvier's beaked whales were found floating at sea or stranded along the coasts with signs of nets on their bodies. These were just the tip of the iceberg of a much greater massacre: an unknown number of carcasses sank to the bottom and disappeared from sight, and could not be added to the tally.

In the late 1980s, the times were changing with regard to public sensitivity to such destruction of wildlife, and the news of the slaughter inundated the Italian media and caused a maelstrom of public indignation that politicians could not ignore. Many advocacy organisations joined together into an ad hoc alliance to petition the government to intervene, collecting more than 150,000 signatures supporting a petition to stop driftnet fishing in Italy. I was among those who physically carried boxes containing these signatures to the Rome office of a rather unsympathetic lady who was then the Minister of Agriculture and Fisheries.

The destructive impact of driftnets was a concern not limited to Italy, nor to the Mediterranean nations for that matter. Worried by the pernicious worldwide development, in 1992 the UN General Assembly adopted a resolution calling for a global moratorium on driftnets. It took no less than another 10 years and countless more cetacean deaths for the European Commission to follow up by banning driftnet deployment throughout the European Member states. Soon, the General Fisheries Commission of the Mediterranean and the International Commission for the Conservation of Atlantic Tuna followed suit with similar decisions, thereby extending the EC-sanctioned ban to the entirety of Mediterranean waters. Rules notwithstanding, driftnet fishing continued undisturbed as an illegal practice in many Mediterranean areas, including in the south of Italy, with lawmakers happy to turn a blind eye—as they still do, albeit no longer with the earlier complete indifference.

During that same period, a budding community of field scientists started to investigate the ecology of cetaceans in the seas around Italy with rather basic but effective scientific tools. That effort had begun to generate data on the distribution and abundance of the different species. Believe it or not, ecological data on Mediterranean cetaceans did not exist before then. At the Tethys Research Institute, we dedicated four summers in the late 1980s to search for cetaceans throughout the Italian seas from Ventimiglia to Trieste, i.e. from the western Ligurian Sea to the North Adriatic. Whales and dolphins were approached to identify the species and determine group size, and data on relative abundance were collected. The results of this effort were as surprising as they were significant: the Ligurian Sea was by far the most important of the seas surrounding the Italian peninsula, both in terms of species diversity and abundance.[32]

---

[32] Notarbartolo di Sciara G., Venturino M.C., Zanardelli M., Bearzi G., Borsani F.J., Cavalloni B. 1993. Cetaceans in the central Mediterranean Sea: distribution and sighting frequencies. Bollettino di Zoologia (Italian Journal of Zoology) 60:131–138. https://doi.org/10.1080/11250009309355800.

A few years later, during the summer of 1992, Greenpeace provided key support for our investigations by making its ship *Sirius* available as a research platform, which allowed the application of line-transect methods to estimate cetacean absolute, rather than relative, abundance. This means that the data collected allowed us not only to compare sighting frequencies among different locations or different times but also to provide actual estimates of population numbers. We crisscrossed the western Ligurian and Corsican seas for weeks and ended up estimating the presence of 900 fin whales and 25,600 striped dolphins in the survey area.[33] We were astonished by such numbers. The support of Greenpeace for the Mediterranean scientific community marked the start of collecting robust cetacean ecological data in the region that no academic or government institution had previously bothered to assemble. The unsung but fundamental contribution of nongovernmental organisations (NGOs) to securing essential scientific knowledge should be recognised, as it allowed for the first time for the development of conservation policies and management measures based on a solid scientific basis.

Admittedly, the notion that the western Ligurian Sea was important for cetaceans should not have been surprising. What was new was the importance of the Ligurian Sea compared to other portions of the Mediterranean. No less than 2000 years ago, the Romans named an area of the western Ligurian coast *Costa Balaenae* (Coast of Whales). More recently, in 1899, Albert I, Prince of Monaco, who alternated his sovereign duties with state-of-the-art oceanographic and marine biological investigations—enshrined in the magnificent oceanographic museum in the Principality—wrote that whilst he had conducted many research campaigns in the Arctic without sighting a single whale, he could see many of them from the windows of his palace in Monaco.[34] That knowledge, however, was not mainstream in our times, and people—including many well-known zoologists—ignored that whales are regular inhabitants of the Mediterranean Sea and specifically of the Ligurian Sea. Still to this day, despite three decades of newspaper articles, TV talk shows, books, social media postings, and popular documentaries, I am often met by general disbelief when I state that at least a thousand whales frolic seasonally in the waters adjacent to the west coast of Liguria.

So, around the early 1990s, on the one hand, we had the news that the Ligurian Sea harboured unexpectedly significant ecological and faunal riches;

[33] Forcada J., Notarbartolo di Sciara G., Fabbri F. 1995. Abundance of fin whales and striped dolphins summering in the Corso-Ligurian Basin. Mammalia 59(1):127–140.

[34] Albert 1 Prince of Monaco. 1899. Sur les animaux bathypélagiques obtenus par la capture des cétacés. Verhandlungen des Siebenten Internationalen Geographen-Kongresses, Berlin. Vol. 2:307–311. Published in Berlin in 1901 by W.H. Kühl.

on the other hand, the awareness had grown of the huge risk that these newly discovered natural values incurred through mismanagement of human activities at sea—fishing with driftnets being just the most glaring example. As explosive as this combination should have been, the bleak situation was instead met by most torpid institutional inertia, leaving me to wonder whether there was any hope that anything could be done.

A wild idea started to take shape in my mind: why not establish a large, international MPA in that part of the Mediterranean? At that time, in my greener days, I was not ready to give adequate consideration to the challenges entailed in such an idea, at a minimum, convincing three nations—France, Italy, and Monaco—to tackle the inevitable political conflicts between the protection of cetaceans and the intense human activities occurring in those waters, such as fishing, shipping and tourism. A far-from-negligible obstacle to the plan was the notion of establishing legal protection over a sea expanse that was beyond the jurisdiction of any nation. In those days, the Mediterranean nations had not yet declared their Exclusive Economic Zones, which can extend to 200 miles from the coast, and national jurisdiction only extended to the 12-mile width of the territorial waters.

In retrospect, today I cannot feel the same irritation I felt when confronted with the perplexed expressions of the lawyers with whom I was discussing the project. To better understand the foolhardiness of creating a protected area in the "high seas"—i.e. beyond the limits of the territorial waters—it is worth remembering that enforcing a protective regime requires an appropriate legal entity to have jurisdiction over the area. Protection involves the ability to pass a law extending such protection and ensuring the law is respected by enforcing it. How can such a goal be attained outside the jurisdiction of any country? The high seas are, by definition, "areas beyond national jurisdiction"; therefore, the conventional legal thinking at the time considered that establishing a protected area in the high seas was utter nonsense.

To me, instead, the real nonsense was the inability to legally protect from human abuse a valuable and vulnerable natural entity—such as a population of whales—only because these whales happened to live in waters beyond imaginary lines of national jurisdiction traced by humans on a map. Such was the case for the fin and sperm whales and their many smaller cetacean relatives in the western Ligurian Sea, living in a marine area falling, to a large extent, in the high seas. If the conventional legal thinking of the time prevented the necessary establishment of protection on the high seas, in my mind, the only path forward was to change the conventional legal thinking. And in fact, 30 years later, this is now happening. Today, the idea of establishing a protected regime over ecologically valuable portions of the world's high seas—encompassing two-thirds of the ocean surface or almost half the whole planet—has finally gained international traction at the UN level, and the entry into force of an international treaty is finally in the home stretch after many years of intense negotiations.[35]

Meanwhile, policy developments in those years were coming to the rescue, removing the notion of establishing a high sea protected area from the realm of romantic daydreamers. Most significant in this regard was the adoption in 1995 by the Mediterranean nations of a revised Barcelona Convention Protocol on Specially Protected Areas (the "SPA Protocol"). This revision foresaw, amongst other things, the possibility of establishing protected areas in the Mediterranean high seas, provided that all parties to the Protocol were in agreement.

Those were the conditions at the beginning of the 1990s when I began to draft a project, in collaboration with Fabio Ausenda, to establish a cetacean sanctuary in the northwest Mediterranean. I named our effort "Project Pelagos".[36] In 1991, a handful of good-hearted gentlemen from the Rotary Club, who supported the idea, formally organised a meeting in Monaco to present the project to the public.

I still remember vividly the day—it was the second of March 1991—when I stood on the podium inside the imposing conference room of the Musée Océanographique, presenting Project Pelagos to a large audience presided over by Monaco's Souverain, Prince Rainier III. As I framed the proposal against the background of our recent scientific discoveries in the very waters lapping against the building's foundations, I could not help but feel the symbolic importance of presenting my ideas to the great-grandson of the man who had built the Museum. I tried to adopt a tone of unfaltering conviction that the notion of creating an international sanctuary in the area was the most evident of imperatives. But I watched with apprehension the thoughtful expression on the face of the old Souverain sitting in front of me.

Eventually, beyond all expectations, things started to fall into place in a string of events that were nothing short of miraculous, given the circumstances. Prince Rainier provided the full weight of his support for the idea. That support caused the concept of a sanctuary to start creating ripples across arenas I could not have dreamed of even days earlier. During a summit of European Union environment ministers in Edinburgh, Carlo Ripa di Meana proposed over breakfast with his French counterpart, Segolène Royal, to start

---

[35] The "Biodiversity Beyond National Jurisdiction" negotiations are described here: https://tinyurl.com/bdf8c96d.

[36] From ancient Greek πέλαγος (pélagos, "sea").

trilateral negotiations about the Pelagos Sanctuary. Back in Italy, Ripa di Meana's diplomatic counsellor, Ambassador Giuseppe Cassini, called for a first meeting with his French and Monegasque counterparts. I was invited to the meeting, and it was there that I first met Professor Tullio Scovazzi from the University of Milan Bicocca, a scholar of international maritime law and the principal expert on the subject, also a consultant to the Italian Foreign Office. As I later learned, Tullio had played a major role in the revisionary work of the Barcelona Convention SPA Protocol and would have a hand in introducing the norm for establishing protected areas in the Mediterranean high seas. Far from being the type of lawyer who thought that designating high seas protected areas was gobbledegook, Tullio bears significant responsibility for transforming mainstream legal thinking regarding protecting the marine environment that occurred at that time.

From then on, plans progressed at a rapid pace. Several meetings in Monaco and Paris followed that first one, and the language of a new treaty began to take shape. By that time, my professional life had also taken a considerable turn. In 1996, a vacancy arose in the presidency of ICRAM, the agency responsible for providing scientific support to national marine conservation policies, and Carlo Ripa di Meana suggested that I be considered for the position. It is a long story, but in 1997 I was eventually appointed to the position by presidential decree. It was a dizzying change from leading a small NGO working with half a dozen colleagues on a shoestring budget, to being placed at the helm of a government body with hundreds of employees and chairing a board composed of the general directors of the relevant ministries. It was a totally new affair having to deal on a daily basis with the mechanisms of national decision-making and to be in direct contact with the wildest collection of ministerial characters. The experience was inspirational, hugely formative and interesting, and it lasted almost 8 years until, nearing the end of my second and last mandate, I was thrown out of office by a new government keen on clearing the table from elements having anything to do with the conservation of nature.

I was in the early years of my position as a civil servant when the day arrived—25 November 1999—when the ministers of France and Monaco were invited by their Italian counterpart, Edo Ronchi, to gather in the exquisite, Raphael-designed meeting room of Villa Madama in Rome for the signing ceremony of the Pelagos Sanctuary treaty. Minister Ronchi asked me to provide the illustrious guests with a sense of the uniqueness of the marine area covered by the treaty, and I broadcasted a recording of sounds emitted by fin whales in the Ligurian Sea. It was a solemn and emotional moment when the loud, low-frequency moans echoing in the grand room caused the tall window panes to vibrate in the otherwise absolute silence. The sound conferred a sense of meaning to the piece of paper the assembled dignitaries were about to sign. Later that day, the agreement having been formalised, the world's first MPA on the high seas was born. In recognition of the idea's origin, the ministers gave the Sanctuary the name "Pelagos".[37]

Pontoporia was now proceeding under the push of her auxiliary engine across the calm waters of the Sanctuary. The sky was overcast, and the night was dark. I knew these waters were teeming with life, although none manifested itself during that night. Progressing at this pace, I anticipated reaching the Italian coast the following afternoon. Sanremo, where the Tethys Research Institute had based its field activities since 1986, thereby creating the scientific basis for establishing the Pelagos Sanctuary, was my final destination and the end of my journey aboard *Pontoporia*.

Day 46: Pelagos Sanctuary for Mediterranean Marine Mammals

I proceeded under engine power over flat-calm seas shortly after a windless dawn. The sunlight barely filtering across the thick cloud cover produced a gloomy, almost oppressive, leaden atmosphere. A strong Mistral was forecast within the next 12 h, eventually sweeping those ugly clouds away. I only had about 40 miles to go and hoped to reach a safe harbour before the wind started to blow in earnest. The day's humid, hazy air was curtailing visibility and hiding the approaching coast of France to the port side. Despite the sunless day, the conditions for sighting marine wildlife were optimal because of the lack of wind, and I was hoping for an encounter to cheer me up during the final hours of my journey.

---

[37] Notarbartolo di Sciara G., Agardy T., Hyrenbach D., Scovazzi T., Van Klaveren P. 2008. The Pelagos Sanctuary for Mediterranean marine mammals. Aquatic Conservation: Marine and Freshwater Ecosystems 18:367–391. https://doi.org/10.1002/aqc.855.

Sure enough, a disturbance at the surface became visible in the distance, slightly to starboard, revealing the presence of something large, maybe a whale or a school of dolphins. I adjusted my course to the right, where the disturbance was last seen, cut the engine, and waited.

Soon, I saw something very big swimming under the surface, leaving ripples in its wake until it finally came out for air. It was a fin whale, no doubt. Still, something was not quite right: the animal was moving too slowly, almost lethargically, and its blow, normally the very manifestation of power, was barely visible, its noise a mere whisper. Once emerged, the whale kept swimming at the surface, almost as though it did not have the strength to arch its back and dive after the characteristic short series of blows. Not only was the whale behaving as if it were exhausted, but it appeared to be severely emaciated, the spinous processes of its vertebrae clearly showing through the skin along its back, dinosaur-like.

I had never seen a cetacean in such poor condition. While I wondered what might have caused this, the whale seemed to summon all its strength to dive, arching its back and pointing its head down as vertically as it could. By doing so, it lifted its tail clear of the water, and everything became horribly evident.

The whale had no tail. It had just a stump with both flukes brutally cut off. With a flash of recollection, I immediately realised that I had seen that whale before. As I mentioned when cruising in the waters south of Crete several days earlier, in 2005 we had encountered a fin whale in the Pelagos Sanctuary that had one of its flukes missing, sliced off the tail, likely in too close an encounter with the keel or the propeller of a ship. We had also noted how this whale, whom we had nicknamed "Mezza-coda" ("Half-tail"), had the peculiar habit of lifting its tail out of the water when diving, something fin whales do not normally do. At that time, we were relieved to note that the animal appeared otherwise healthy, indicating that it was managing to live a seemingly normal life despite its disability, including being able to dive and feed.

"Mezza-coda" was a well-identified individual in Margherita's fin whale catalogue based on its unique disability as well as for its distinctive, light grey pattern on its back (used to identify fin whales like a fingerprint for humans), including a chevron-shaped shading behind the blowhole and an asymmetrical blaze on the right side of the nuchal region. However, I now saw with dismay that the poor whale had incurred a second, this time fatal, misadventure. Either "Mezza-coda" had been hit again by a ship, or a fishing line or a piece of net caught around the remaining fluke ate into the flesh and eventually sliced it away altogether, transforming the whale from "Mezza-coda" to "Coda-mozza" ("Docked-tail"). Weakly, slowly, probably painfully, but forward it went, evidently without the strength needed to dive to the depths where it could find its prey. I was astonished to see the poor animal managing to move at all, given that the tail is a whale's means of propulsion.[38]

That encounter overwhelmed me with a crushing sense of exhaustion, and I felt such a heavy sadness descend on me that I wished for a moment to sink to the bottom of the sea and stay there. What hope could we have to make a difference if even the largest, strongest animals on the planet could not survive our influence in the immenseness of the sea? Even here, in a sanctuary that had been declared with bells and whistles specifically for their safety and well-being? Did any of my efforts, and those of so many others like me, invested in maintaining the Mediterranean marine environment, make any sense in the face of such a blatant demonstration of failure?

It was not just the deep sadness I felt for the fate of the poor individual whale, the pain it had to endure, its condemnation to starvation for the sole sin of existing in an environment made so inhospitable by us humans. My frustration went well beyond that. What about all our efforts to protect this area if we cannot allow its inhabitants to live a decent life? Even assuming that we manage to ensure the survival of the whale population here—in itself not a given at this moment in time—what is the point of mere survival if it means a constant struggle to avoid entanglement in a net, being chopped up by a propeller, being driven away by the loud seismic surveys used to search for oil and gas in the sea floor, or being intoxicated by poisonous chemicals or microplastics?

Noting that the conservation effectiveness of the Pelagos Sanctuary was much lower than what I had expected, insofar as the conservation of cetaceans was concerned, was aggravating. How would the whales reply, had I the power to ask them whether they had noticed any difference in their lives from the day the Pelagos Sanctuary was established? Better to be unable to ask this question, after all. In the years following its declaration, the Pelagos Sanctuary became well-known as an "unconventional" MPA, even internationally, and was also inscribed in the list of "Specially Protected Areas of Mediterranean Importance" (SPAMI) of the Barcelona Convention. But even this did not imply the decision of endowing Pelagos with proper management.

In my original proposal, the Pelagos Sanctuary should have been managed under UNESCO's Man and the Biosphere Programme as a Biosphere

---

[38] I wish the story of "Codamozza" were a fictional episode that I invented to end my story with a bang of drama. Unfortunately, it is not. More details on the unfortunate whale can be found here: https://tinyurl. com/3utj3ud5 and https://tinyurl.com/2nxfpu8j. First sighted with a single fluke in the Pelagos Sanctuary in 2005, It was seen again a few years ago off Spain and France with both flukes gone. After that, the poor animal was repeatedly observed following an itinerary which is rather incredible considering its handicap: first off Syria, then Greece, Sicily and Calabria, and finally in the Pelagos Sanctuary again, where it was seen for the last time.

Reserve.[39] The programme would have ensured the establishment of a management body responsible for its good functioning. Biosphere Reserves also present the opportunity to have an umbrella framework for cooperation and would bring international recognition, support for marine and coastal monitoring, research, education, and management, and foster—among other things—economic development and expanded livelihood opportunities. Unfortunately, at the Conference of Parties of the Pelagos Agreement which was held in Porquerolles in 2006, France inaccurately argued that the Biosphere Reserve model cannot apply to the Pelagos Sanctuary because these reserves are supposed to only be land-based, and that ended the discussions. The parties had opted instead to adopt management plans without the support of a dedicated management body, significantly weakening the potential of effectively managing the area for conservation. Admittedly, granting effective protection to marine life over such a wide area, and one where human activities are so densely concentrated, is no mean feat—and there is a limit to what management action can accomplish short of banning most human activities from the area, an obviously unthinkable proposition. However, and exactly because of the challenges to conservation imposed by such a bold political act, an extra effort to achieve management effectiveness was expected and hoped for.

Establishing the Pelagos Sanctuary did achieve one important outcome, however. The coastal communities bordering the Sanctuary, from the largest of cities to the smallest towns, were given the opportunity to establish a partnership in spirit with Pelagos. Through a formal arrangement with the relevant Ministry of the environment—France, Italy, or Monaco—these communities now fly the Sanctuary's flag in a symbolic allegiance to the treaty and its inherent purpose. This simple initiative created widespread awareness among the people of the existence of the animals and their plight, and an ideal emotional link between these communities and the cetaceans living in Sanctuary waters. Nowhere this process of getting a local community excited about being part of Pelagos has worked more effectively than in the area involving Sanremo and the surrounding municipalities, thanks to the passionate involvement and activism of colleague and friend Sabina Airoldi, whom many today consider the Sanctuary's mummy.

The struggle to get Pelagos to function and achieve its goals must continue despite all setbacks, as it must continue everywhere else throughout the world's oceans. Overfishing, illegal fishing, marine traffic, pollution from toxic chemicals, plastic, and noise are the many Horsemen of a Marine Apocalypse that

---

[39] UNESCO's Biosphere Reserves: https://en.unesco.org/biosphere.

we must fight. We have no other choice, despite the meagre results, despite our occasional despair. We cannot lose hope of eventually making a difference—and in the end, we may succeed.

Good things are happening in the Mediterranean thanks to the efforts of women and men of goodwill. The numbers of bluefin tuna have increased in recent years since the management of the fishery has become more effective and less corrupt. Monk seals are showing signs of a new ability to reconquer their lost terrain, perhaps due to changes in human attitudes towards them, partly influenced by awareness campaigns. Loggerhead turtle numbers have also increased, and nesting sites are multiplying across the Mediterranean. Bottlenose and striped dolphins seem to be doing better today than in the recent past, the latter having recovered from a devastating morbillivirus epidemic in the 1990s—the immunodepression they were affected by having been enhanced by chemical pollutants now discharged at sea in lesser quantities. These are all encouraging signals that persistence in the struggle to improve the condition of wild species can bear fruit. Furthermore, many of these species can act as umbrellas, extending their protection benefits to less charismatic critters who happen to share their environment with them.

We should be careful, however, to place these isolated successes within the greater, planet-wide picture to avoid risking unwarranted optimism that would prevent us from clearly seeing the impending planetary catastrophe that our actions are causing. What would be the point of painstakingly restoring a masterpiece painting to its original splendour, if the museum in which it is conserved risks to burn to the ground? As we restore the individual paintings, we must also ensure that the collection is protected against the risk of fire. Unfortunately, we are far from such protection for the state of the oceans. And yet, the situation is entirely in our hands. All the wounds currently impairing the health of the Mediterranean can heal, and it is we humans who can heal them. We can do it, and therefore we must do it. Who other than us humans can slay the monster, given that the monster is us?

The weather changed entirely around noon. The sky to the west had cleared of the clouds, and the first foreboding gusts of Mistral had made themselves felt, causing *Pontoporia* to heel over slightly and jump forward with a sprint and a faint creak of her rigging. Under the clear sky, the colours of the world were restored as by magic, the sea had regained its dark blue appearance, and pretty little white breakers had started to appear around the boat. In normal circumstances, I would have raised the sails to celebrate the change of mood and take advantage of the wind's push, but it was no longer worth the effort since the port of Sanremo had already come into view.

Sanremo had been the base of Tethys' research campaigns in the Pelagos Sanctuary for the past three decades. Since the early days of our research activities, our boats have been benefiting from the hospitality of the Marina of Portosole, thanks to a significant extent to the interest of the marina's historical director, Captain Pierfranco Gavagnin, an old salt who had an incoercible passion for whales. Given our history, I felt that it was fitting to end my travel here.

Thirty minutes later, having sailed across a choppy sea that was mounting fast under the whip of a lively Mistral, the boat was safely docked inside the marina. My journey aboard *Pontoporia* had come to an end, but the journey of all men and women of goodwill who wish to fight for the oceans is only beginning.

The journey is a "long one, full of adventure, full of discovery", just like Constantine Cavafy would have prescribed.

10

Epilogue: Patmos

The big ferry had left the Piraeus well after midnight, with some delay due to the large number of cars and trucks that had to be squeezed into her hold. Having spent some time in the hot Attican night watching the embarking operations from the top of the ferry's highest stern bridge, I retired with Flavia to the quiet of our cool cabin, considering how things had improved during recent decades in Greece as far as ferries are concerned, and how ferries could be taken as an example of the nation's changes under many respects. The transition from the crowded, noisy, and smoke-filled environment one had to endure for endless crossings aboard the archetype of the twentieth-century Greek ferry to the current condition felt like we had been promoted from hell to heaven.

The passage across the Aegean was a bit rough due to Meltemi, but unlike *Pontoporia*, the ferry was large and mighty. The vessel's mildly bumpy gait had not affected my sleep, and delegating the progress across the sea (not to mention my safety) into the hands of the captain of a big ship was a welcome change. After so many days spent on the water with only myself to resort to for help, my craving for self-determination had waned considerably.

Shortly after dawn, the ferry made a stopover on the island of Ámorgos, and the loud metal-on-metal clang of the anchor chain descending through the hawse-hole during the docking manoeuvre in the harbour of Katàpola had woken me up. After that, I could no longer catch sleep. I had travelled to Patmos countless times over more than half a century, yet getting there was invariably a reason for excitement.

The cabin had a window overlooking the ferry's bow, which allowed me to see where we were going. The ferry was still 40 miles away from Patmos: all I

G. Notarbartolo di Sciara, *Sailing Across a Wounded Sea*,
https://doi.org/10.1007/978-3-031-54597-9_10

could see in front of me was an empty sea and a sky festooned with a line of tiny cumulus clouds hanging above the horizon. Soon after, however, as the ship advanced, one of the small clouds appeared to stand out, stationary instead of moving in concert with the others; it was also shining in a different shade of white. Although it was too far to make out the distant land's profile in the morning haze, I knew that that little speck in the air was not a cloud. It was the Chora of Patmos, the village of whitewashed buildings huddled around the ancient monastery of St. John the Theologian on the top of the hill. Celestially suspended in the sky by virtue of an optical trick induced by the distance—but in a fitting metaphor of its metaphysical quality. Most importantly, it was home, and it was where I was going to be for the foreseeable future.

Two hours later, having docked in the Patmos' port of Skala, the ferry opened her stern and dumped onto the quay her multitude of passengers gleefully running out and away, like many Pinocchios stampeding out of the whale's belly. Two of those stampeders were me and Flavia, mad keen on making our way up the hill and reaching the Chora, our destination at the end of a long journey.

Friends have often asked me why I have such a fondness for this place. Patmos is just one amongst hundreds of inhabited Greek islands, all endowed with their special charm. Her hills are greenish-brown in summer and bright green, strewn with fragrant blankets of flowers in winter and spring, like most Greek islands. Her coastline is deeply indented, creating breathtaking landscapes with alternating bays and inlets lined in their innermost parts with small, secluded beaches. However, this is also like so many other Greek islands. Even the village of Chora on top of the hill—a stunning agglomeration of immaculate dwellings huddled around an imposing 1000-year-old monastery—is not that unique amongst Greek islands despite having been awarded World Heritage status by UNESCO. The right answer is that it is impossible to rationalise fondness for a place because of the prevalence of irrational elements responsible for such fondness. I happened to set foot on the island half a century ago by chance and became seduced at once by the island's powerful *genius loci*, convinced—as I still am—that Patmos is the most beautiful place on Earth. Having visited hundreds of islands during that nomadic phase of my life, once I got to know Patmos, I felt I did not need to look any further when the time came for me to be keener on a more sedentary lifestyle.

Something intriguing occurs to mind about falling in love with a place. I see this as being very much like falling in love with a person, as if humans had a standard way of falling in love that they then apply to different entities, as appropriate. Before contact, one's imagination is ignited by the exotic magic of the site, starting from the sound of its name (*Patmos* in this case: what could have been more intriguing already?). The idea of the site then starts building up in the imagination, nurtured by bits of information picked here and there, in loose correlation with reality. Having finally reached the place and having had the chance to explore it, one becomes enamoured by its pervading spirit and the many ways it surrounds one's senses. The goodness of choice as a place to stay becomes consolidated conviction, and any divergence from the previously imagined idea is forgotten as irrelevant because reality is better than imagination. Until then, the early magic is still very much alive. With time, however, knowledge creates habit, and habit replaces novelty, nudging that wonderful original sense of enchantment into the background. Finally, one decides to embrace the place, dive into it, settle into it, take up residence in it, and become one with it; the sense of ownership becomes dominant, and the site gains immensely in importance. Alas, at that point, the benefit is blemished by a new cost because the original charm is gone; ephemeral whiffs of it may come back occasionally, the old sentiment evoked by the notes of a musical instrument or by the smell of a dish tasted only there, but there remains in the background the sense of regret for the disappearance of those early tingling sensations and for the loss of that blissful state of endearment that—as we know too well—will never come back. Knowledge trumps romanticism in an obligate direction like a weird doppelgänger of entropy, and all we can do is let ourselves be dominated by the process.

We opened the door of the old house and walked into her shaded embrace, which was shielding us from the outside heat and its blinding light. The whitewashed stone walls, ridiculously thick, created a protective indoor habitat, and a wonderfully cool air was causing the window curtains to flutter softly under the breeze. A faint smell of dampness preserved from the springtime rains created a stark but pleasing contrast with the parched world outside. A tray full of ripe figs and a bowl of *xynomyzithra* goat cheese were waiting for us on the marble table in the kitchen, a gift from the land left there by our neighbour Grigoritsa when she learned of our imminent arrival.

I come to Patmos to dedicate time to myself, to recharge my batteries, to catch my breath from a life running too fast for comfort, to recover the opportunity of giving things their forgotten importance and spend time with the people I love, or simply like. Mount *Profitis Ilias,* the island's tallest peak, stands in front of me across the drawing-room window like a guard to shield the peace of my mind, and the azure expanse of the sea beyond the mountain to the west reassures me that humanity's pandemonium is too far away to be in a condition to disrupt such peace. The protecting landscape is conducive to positive thoughts, calm disposition, and creative inspiration. And yet, the tendency to worry about the degradation of the planet's conditions, recognisable wherever I lay my eyes, is so ingrained in my way of being that I find it impossible to avoid contemplating Patmos more as an entity in need of healing than as one in charge of healing me.

As I got deeper and deeper into observing the status of the Patmian environment, I discovered with surprise how close the local conditions and challenges replicate, on a smaller scale, those of the wider Mediterranean, if not even those of the planet. The first observation that hit me somewhat brutally was a striking sense of contrast.

On the one hand, the pureness of the landscape, typified by an exquisite balance between the natural elements and the changes to such landscape elegantly introduced by the island's human hand across the millennia: the parched hills disseminated with sparse, whitewashed, exquisitely built farmhouses and tiny chapels, laboriously terraced across centuries by a workforce emigrated long ago to distant continents. On the other hand, the sense of loss one feels watching such balance disrupted in places where the built-up areas have become dominant during the past decade or so, with the accruing of constructions characterised by the disappearance of the good taste of lore. Whenever, to some extent, this has happened because of the increased needs of an expanding local community, I do not dare deny the right to satisfy such needs. More often, though, the encroaching of the built-up surface onto the natural space is instead the result of the high real estate value recently acquired by Patmian land, as the island had become fashionable in the international arena. This circumstance makes Patmos prey to non-native developers, false prophets of a god of progress, circling like vultures to offer promises of wealth to locals where the only wealth produced is ultimately their own. As a result, the risk is real that Patmos will be progressively but permanently stripped of her unique aesthetic richness—and her children deprived of their most valuable heritage. Keeping the balance in the landscape between built-up and natural areas is first and foremost in the interest of the local inhabitants as one of the main reasons for the island's attractiveness; the matter has significant economic implications. Striking some balance between natural and human-modified landscapes is not a unique challenge to Patmos. It is a condition replicated at all scales, all the way up to the global need to allow nature the space it requires to maintain its functions and diversity—as humanity devours the land and encroaches on the planet's surface for its purported urban, agricultural, and industrial needs.

As one moves from landscape to seascape, the situation worsens considerably. When sailing *Pontoporia* south across the Dodecanese, I noted how the underwater world surrounding these islands—at least as far as the infralittoral zone is concerned, down to a few tens of metres of depth from the water line—is startlingly barren. Whilst seagrass meadows are still doing reasonably well in many places, the cover by the true algae, where one still exists, consists of only a handful of predominant species, giving the seascape the dull, uninteresting aspect of an underwater monoculture. The classic representatives of the Mediterranean benthic fauna, such as the bright red starfishes, the sea urchins, the sea cucumbers, as well as various shelled molluscs, crabs, and shrimps, have become so rare that they are hard to spot. Octopuses are wary about coming out of their holes. Fishes also occur sparsely, most of them tiny—as if they had not had a chance of growing to adult size—with native species yielding to alien invaders arriving via the Suez Canal.

Marine mammals such as bottlenose and common dolphins, often found frolicking out in the open sea surrounding the islands, keep clear of the Patmos coastal waters, wary of unpleasant encounters with the fisherfolk. Even monk seals are seen here only on rare occasions, despite the presence of a perfectly suitable habitat for their kind. The general impression is one of a dysfunctional ecosystem, incapable of sustaining a healthy balance of life made of the usual consortium of producers, herbivores, predators, and decomposers—and this includes the artisanal fishers, who are at least in part responsible for this condition of degradation, and at the same time amongst its main victims.

As I described in greater detail during the Aegean passage of the journey of *Pontoporia*, marine ecologists are still unable to detect a single smoking gun to explain the unhealthy state of the coastal marine environment of the Dodecanese. This has proven to be elusive because the smoking guns are likely to be more than one. Fishing in the first place, with its rogue siblings overfishing and the illegal use of dynamite, accompanied by sea warming, alien invaders from the Red Sea and the tropical Atlantic, and poor water quality deriving from untreated urban sewage discharged at sea. All these agents, taken together, concur to inflict multiple whammies onto a marine environment that is weak from the start due to the inherent low productivity levels of the southern Aegean waters. These factors are not exclusive to the Patmian seas: fishing, marine traffic, pollution, invasions, and global warming all add to the many wounds from which the wider Mediterranean suffers. And not only the Mediterranean, for that matter, given the worldwide relevance of these factors. These are perfect examples of how environmental problems at the local scale are replicated at the ocean scale, and ways of addressing them and mitigating them seem to be equally formidable despite the different levels at which they occur.

One afternoon I decided to take to the sea with my caique to spend time off the south coast of Patmos. That part of the island is not served by roads and can be reached only by boat, and the place is still beautifully pristine, with hardly a single scar having been inflicted on the landscape by humans. In the large *Prassovounò* bay, the high, reddish cliffs slope down steeply to the water-line, covered in places with vast patches of evergreen lentisk and rock rose in a pleasing contrast of colours. I anchored the caique in a sheltered nook of the bay and took a quick swim in the cool, dark blue waters. Soon after, comfortably stretched on the bow deck, I was negotiating through the pages of Laozi's *Tao Te Ching*, enthralled by the guidance conveyed by the ancient text prescribing readers 2400 years ago to search for their lost relationship with the natural world. As I progressed in my reading, I wondered how little human nature had changed across the millennia, whilst still admiring how inspirational and of unexpected modern relevance was the book's underlying notion that all living beings are tightly interrelated. However, the thickness of some passages was conjuring with the gentle rocking of the boat and the warmth of the afternoon to send me rolling into waves of drowsiness I had difficulty controlling.

In that condition, a lapping noise from the sea below me caught my attention. Leaning over the gunwale, I discovered with surprise that none less than a monk seal was staring at me from the sea surface, right under the bow of the caique. Not only that, it was the same seal I had encountered in that very place two decades before, recognisable from that distinctive white spot she had between her eyes.

She was now an old animal who had seen lots of change and managed to survive through it, and the signs of that wisdom were visible on her aged face and scarred body. And then, even more astonishingly, the seal spoke to me.

I have been running away from you. All my life, I have been hiding from your murderous and destructive presence. In the end, my tribe is now giving signs of recovery, but that is perhaps because you have moved your attention from the effort of catching the last fish in the sea to more profitable ventures on land.

From my hiding, I have been observing you humans since our last encounter so many years ago, and I could see that instead of making progress in your relationship with the natural world, marine or otherwise, you have gone backwards. Because of your actions, the sea I am swimming in has never been so warm in my recollection, and the fishes that I now find here I had never seen when I was younger.

Your contraptions are belching greenhouse gases in the air increasingly year after year despite your stated intentions. Now that the effects of climate disruption deriving from that misbehaviour are touching you in your soft belly, you are getting excited, and you forget that climate disruption is just one of the many damages caused by your recklessness: what about the waves of animal and plant species extinction? The sad irony is that you will be punished by these losses like everybody else. But you don't seem to care or even understand.

Your conservation thinking is full of jokes, "sustainable development" being the most ludicrous of them all: how could you think it can be possible to keep growing within a finite space – because that is what the planet is - and being sustainable at the same time? You say that nature is to be conserved so that it keeps providing you with "ecosystem services", but if those services are only for you humans, you and your domesticated animals will eventually end up being the sole living beings on the planet, and good luck with that. In a dwindling natural world, if things keep going like now, you will end up being left with the last existing forest confined to a flower pot. You live in the present with rules from the past, and you show no concern for the future of your children and their children.

So, where is this superiority of yours? You think you are so clever because humans have composed symphonies, formulated creative philosophical dissertations and sent people to the moon. In the end, however, your technology is getting out of hand, and your brilliant ethical reasonings are not helping a bit because your greed trumps every other consideration. But with you, as collateral damage, scores of other guiltless beings will be pushed off the cliff.

Instead of feeling superior and detached from the natural world with your self-attributed and delusional divine gift of mastering it, the predominant feeling towards the world you are part of should be love. Your efforts should not be solely dedicated to your well-being, but to the well-being of the whole you are part of – the land, the sea, the air, and, of course, the animals and the plants - because only that can save you. How can you not see the moral obligation you have to buffer the destructive potential you have acquired through your cultural

evolution by treating the planet and the species inhabiting it with the stewardship and respect they desperately require?

Shed the delusion that you humans are separate entities from the rest of the
world: it is a false and dangerous tale. You are an integral part of nature, devoid
of any divine right to use nature's goods and services without consideration for
the effects of your doings.

Forget the wish to conquer nature because you are part of nature - and if there
is anything out there for you to conquer, it is yourself.

As she pronounced those last words, the seal lowered her head under the
surface and, with a strong push of her rear fins, arched her back and disappeared in a swirl of water.

Dumbfounded, my eyes now tried to focus again on the *Tao Te Ching*
volume I had dropped on the deck. But the seal's words were still ringing in
my ears. As I considered that fact, the notion's absurdity hit me: how could
a monk seal be talking to me? I must have been dreaming. The seal's mono-
logue must have come from my conscience, and there cannot be any other
explanation. Maybe there was not even a seal around the caique, no less a
talking one; monk seals are so rare in Patmos nowadays. Even as a dream,
however, the harsh sermon I had just been served, likely through a trick of
the mind, was a telling parable conducive to considering humanity's
predicament.

Such a predicament is so hard to accept. Everyone can see that we are forcing ourselves into a dead end. With ecosystem functioning impaired and an
increasingly significant portion of the planet's biodiversity disappearing,
including in an ocean that we can no longer perceive as infinite, it is our own
world that disappears—including all the immaterial benefits we derive from
it, such as sense of beauty, inspiration, and spirituality. And yet, no effective
action is being taken to do anything consequential about this tragedy. Some
weak signals of concern for the climate crisis exist, with the world's leaders
regularly meeting to vainly discuss actions to stem climate disruption. Still, so
far, the labouring mountain hasn't been able to bring forth even the proverbial mouse.

And this is about climate change, reflected in changes in our human habitat that already affect our well-being in an increasing number of locations worldwide. But what about the extinctions we are causing of the multitudes of other living beings? The loss of biodiversity can have far-reaching impacts on ecosystems and human societies, including decreased food security, reduced resilience to climate change, and the loss of associated cultural and spiritual values. Ironically, mainstream economic thinking today has concluded without any room for doubt that excluding the value of nature from economic valuation is a tragic mistake. In a well-known report commissioned in 2019 by the British government, Partha Dasgupta from the University of Cambridge has eloquently described nature as humans' most precious asset, which is being squandered in a devastating sequence of mismanagement blunders.[1]

The conservation challenges facing humanity are not impervious to solutions. Yet, objective observation will conclude that efforts to redress all the wrongs concerning the relationship of humans with their environment—the Mediterranean Sea being an excellent example—are weak, incoherent, and ultimately insufficient. There are many reasons for this, including, most notably, choices in the short-term deeply rooted in the cultural inadequacy of decision makers, favouring business interests that clash with sustainable behaviour. Such demeanour is oblivious to the longer-term collective interest, in particular those of the future generations. Ultimately, I think that the primary disease affecting humans, as far as their relationship with the planet is concerned, can be reduced to a lack of a vision regarding where humanity wishes to go, and the failure to understand the absolute necessity of having such a vision.

The need for humanity to have a vision of where it wishes to go is a novel concept. *Homo sapiens* is one of many animal species on the planet, and animal species typically do not have such a requirement. Do sea urchins need to have a vision of where their tribe wishes to go in the future? A question I cannot ask because sea urchins have not told me, but I strongly doubt it: sea urchins live their life within their ecological constraints without much fuss. Of all the world's animals, however, humans are an exception because they have managed, through their unique cultural evolution supported by technology, to escape the natural laws that generally regulate the role of species within their ecosystems. This is an extraordinary freedom we have granted ourselves, but also one that does not come for free. This freedom must come with the

---

[1] Final Report of the Independent Review on the Economics of Biodiversity led by Professor Sir Partha Dasgupta https://tinyurl.com/2ud4re8a.

obligation for our species to recognise its responsibilities not only for its own well-being, but for the well-being of the whole planet. An obligation that requires the exercise of restraint, and has both ethical and practical implications. If humans do not restrain themselves and manage their power, they will likely be doomed by their large brains—a trait thereby converted from our species' main asset to a seriously maladaptive one.

The tragedy of our time is that humanity ignores its obligation to restrain. It behaves in ways that are ethically indistinguishable from the behaviour of sea urchins. It insists on carrying on with its insouciant "business as usual" lifestyle as if nothing is happening and the planet's resources are infinite.

This lack of vision is just as evident considering the ineffectiveness of actions concerning decisions taken in the global arena—about measures to tackle climate disruption, curb ocean depredation by large-scale industrial fisheries, or stop polluting the ocean with millions of tonnes of plastic waste— just as it is when decisions are made at the opposite end of the governance spectrum. For example, in the city council meeting rooms of the tiny Mediterranean communities, where discussions concern the management of their lands' meagre fresh water supply or their sewage treatment plant, the containment of the ruinous effects on land texture caused by multitudes of free-ranging goats, or the preservation of the islands' fragile but ecologically essential coastal wetlands.

Is there any hope that the Mediterranean's wounds will be allowed to heal? Are we humans capable of deciding on the future of our sea and, consequently, on our future? Some people are convinced that humanity will be able to continue with "business as usual" and still manage to keep an upper hand over nature by adopting increasingly clever technological fixes, such as extracting more food from land by improving agricultural techniques, e.g. through genetically modified crops, or geoengineering the atmosphere to fight climate disruption by blocking the radiation from the sun. I see these as nothing more than expedients amenable to shift the goalposts further into some time in the future, grounded in a hubristic attitude that can only get us into deeper trouble.

A more radical change is needed, and one that goes far beyond the enterprise of healing the Mediterranean Sea's wounds. No action will be effective if humanity persists with the false conviction of being an entity separated from the natural world, with its purported divine mandate to conquer, dominate, and appropriate as it wishes. There is no room in these pages for an in-depth dissertation of the philosophical bases to critique this attitude, which the interested reader can find by perusing a wealth of more appropriate sources.[2] But I cannot avoid noting that Western thinking started walking on a dangerous path quite early, and with none less than that extraordinary thinker of Plato, who laid down the foundations of a dualistic universe separating reason from body and, consequently, humans from nature. The Church Fathers of early Christianity picked up this concept, which was later transferred onto mainstream scientific thinking at the turn of the sixteenth century by the likes of Francis Bacon and René Descartes.

Their dogma, seen today under the light of the multifarious effects of the actions of humans on the planet and its inhabitants, is by now not only dated: it is tragically wrong. The exclusive attribution of sacredness to an invisible, omnipotent entity above us has distracted humanity from conferring the much-needed reverence to the land and sea that nurture us and that we are part of, which are the true guarantors of our survival and the conduit to our well-being and happiness. The fundamental chasm that the dualistic doctrine has created between humans and the natural world has justified, even sanctified, an attitude ultimately conducive to getting humanity into deeper and deeper trouble. Far from intending to belittle the sophisticated reasonings of those ancient thinkers, I contend that their ideas have failed to withstand the test of time. Their purported relevance has faded, and the gold they might have produced in their days has now turned to poison.

The radical philosophical and societal change involved in having humans freeing themselves from the yoke of a dualistic ideology and behaving as an integral part of nature will require a revolution of epochal dimensions. A change mothers should prepare their babies for when they are still in their wombs.

A deep revision of our entire economic dogma is needed. Such transformation will demand adopting an alternative to perpetual growth, involving the need to make radical choices about the continuation of capitalism,[3]—something the world does not seem ready for yet. Barring violent upheaval triggered by some catastrophic event, such a massive change—provided humans will eventually decide to go this way—will take generations, and there is no sign that this might be happening any time soon. To exacerbate the task, fierce opposition to such metamorphosis can be expected from the powerful

---

[2] Amongst the many works on this subject, I particularly recommend Lent J. 2017. The patterning instinct: the cultural history of humanity's search for meaning. Prometheus Books, New York, and Lent J. 2021. The web of meaning: integrating science and traditional wisdom to find our place in the Universe. Profile Books, London.

[3] Hickel J. 2020. Less is more: how degrowth will save the world. Penguin Random House, UK.

economic forces profiting from the status quo—with the full support of the world's leaders, whether democratically elected or autocratic.

Meanwhile, as the clock ticks, the planet's condition deteriorates under our eyes, making interventions not only increasingly urgent but also increasingly challenging. But let us make no mistake: we cannot afford to be deterred by the challenge, immense as it is.

Besides, resetting the human mind, as daunting as it might be, is surely not as daunting as dealing with exogenous catastrophic natural events such as a meteorite hitting the Earth, a tsunami washing our civilisation into the sea, an earthquake reducing our houses to crumbles, a volcanic eruption annihilating us in a purple cloud of poisonous gas.

As I was considering on the last day of *Pontoporia's* journey when I approached the final destination of the port of Sanremo, who else can succeed in the task of changing the human mind if not us, the humans?

Success will be hard to reach but not impossible, but to succeed, we must try. And as we try, we must not lose the hope that, by trying, we will succeed.

The sun is about to set, and it is time for me to bring the caique back to her mooring for the night. It had been a great day out on the sea, and the warm light of the dying day was dressing the hills with a colour of intense beauty. The figs and the *xynomyzithra* are waiting for me on the kitchen table, and the sensation of being alive in a world I love so much is exhilarating.

I expect many more great days like this to come to preserve a status of happiness throughout the remaining time of my existence. But I cannot accept that such a sense of exhilaration might be denied to future Earth dwellers, humans and non-humans alike.

This consideration alone explains and justifies the continued struggle to set things right back again.

Glossary

AIS Automatic Identification System, an automatic tracking system that uses transceivers on ships to broadcast information about their unique identification, position, course, and speed. On the receiver end, such information can be displayed on electronic charts thereby allowing other ships as well as maritime authorities to track and monitor vessel movements in a given area.

Auxiliary engine In sailing vessels, the auxiliary engine is a secondary source of propulsion which is, however, very useful when there is no wind, and in manoeuvres in harbours.

Charismatic megafauna A category of large animal species, often faring in a precarious state of conservation, that have symbolic value and widespread popular appeal. Being liked by the broad public, they can help focus general interest in conservation, thereby supporting environmental protection.

Full and by Sailing with all sails full and lying as near the wind as possible.

FRA Fisheries Restricted Area, a geographically defined area in which specific fishing activities are temporarily or permanently banned or restricted by the GFCM to improve the exploitation patterns and conservation of specific stocks as well as of habitats and deep-sea ecosystems.

Genoa A large jib or staysail that extends past the mast.

GFCM General Fisheries Commission for the Mediterranean, a regional fisheries management organization that plays a critical role in fisheries governance, and has the authority to make binding recommendations for fisheries conservation and management and for aquaculture development.

GPS Global Positioning System, a satellite-based radio navigation system owned by the United States government and operated by the US Space Force. GPS provides geolocation to receivers anywhere on the planet where there is an unobstructed line of sight to four or more *ad hoc* satellites.

© The Author(s), under exclusive license to Springer Nature Switzerland AG 2024
G. Notarbartolo di Sciara, *Sailing Across a Wounded Sea*,
https://doi.org/10.1007/978-3-031-54597-9

Gunwale The top edge of the hull of a ship or boat.

IMMA Important Marine Mammal Area, discrete portions of habitat, important to marine mammal species, that have the potential to be delineated and managed for conservation. IMMAs consist of areas that may merit place-based protection and/or monitoring.

IMO International Maritime Organisation, the United Nations specialised agency with responsibility for the safety and security of shipping and the prevention of marine and atmospheric pollution by ships.

IUCN International Union for the Conservation of Nature, an international NGO comprising many governmental and non-governmental organisations around the world, provides support to nature conservation including tools such as the Red List of Threatened Species, the Key Biodiversity Areas, and the Important Marine Mammal Areas.

Knot (unit) Unit of speed equal to one nautical mile per hour, exactly 1.852 km/h.

Lessepsian migration The migration of marine species across the Suez Canal, from the Red Sea to the Mediterranean Sea. It is named after Ferdinand de Lesseps, the French diplomat in charge of the canal's construction.

Maestrale, Mistral A wind blowing in the Mediterranean from the northwest.

Meltemi Strong, dry north wind of the Aegean Sea, blowing from about mid-May to mid-September. It is a dominant weather influence in the Aegean.

MPA Marine Protected Area.

Nautical mile Unit of length used in marine navigation, defined as the meridian arc length corresponding to 1 min of latitude at the equator is 1852 m long. The derived unit of speed is the knot, one nautical mile per hour.

NGO Non-Governmental Organisation.

Port (nautical term) Referring to the left side of the vessel, when aboard and facing the bow (front).

PSSA Particularly Sensible Sea Area, an area needing special protection through action by the International Maritime Organisation because of its significance for recognized ecological or socio-economic or scientific reasons and which may be vulnerable to damage by international maritime activities.

Sirocco A wind blowing in the Mediterranean from the southeast.

Starboard Referring to the right side of the vessel, when aboard and facing the bow (front).

Sympatric Animals or plant species or populations occurring within the same or over-lapping geographical areas.

Tethys (organisation) Tethys Research Institute, a research NGO founded in Milano, Italy, in 1986.

Index

© The Author(s), under exclusive license to Springer Nature Switzerland AG 2024   **249**
G. Notarbartolo di Sciara, *Sailing Across a Wounded Sea*,
https://doi.org/10.1007/978-3-031-54597-9